Wood
Preservation

OTHER TITLES FROM E & FN SPON

Defects and Deterioration in Buildings
B.A. Richardson

Building Failures
W.H. Ransom

Building Services Engineering
D.V. Chadderton

Clays Handbook of Environmental Health
Sixteenth edition
W.H. Bassett

Practical Timber Formwork
J.B. Peters

Timber Structures
E.C. Harris and J.J Stalnaker

Timber Engineering
Practical Design Studies
E.N. Carmichael

The Maintenance of Brick and Stone Masonry Structures
A.M. Sowden

For more information about these and other titles please contact:
The Promotion Department, E & FN Spon, 2–6 Boundary Row, London,
SE1 8HN

Wood Preservation

Second edition

Barry A. Richardson

Consulting and Research Scientist
Director, Penarth Research International
Limited

E & FN SPON
An Imprint of Chapman & Hall
London · Glasgow · New York · Tokyo · Melbourne · Madras

Published by
E & FN Spon, an imprint of Chapman & Hall, 2–6 Boundary Row,
London SE1 8HN

Chapman & Hall, 2–6 Boundary Row, London SE1 8HN, UK

Blackie Academic & Professional, Wester Cleddens Road, Bishopbriggs, Glasgow G64 2NZ, UK

Chapman & Hall Inc., 29 West 35th Street, New York NY10001, USA

Chapman & Hall Japan, Thomson Publishing Japan, Hirakawacho Nemoto Building, 6F, 1–7–11 Hirakawa-cho, Chiyoda-ku, Tokyo 102, Japan

Chapman & Hall Australia, Thomas Nelson Australia, 102 Dodds Street, South Melbourne, Victoria 3205, Australia

Chapman & Hall India, R. Seshadri, 32 Second Main Road, CIT East, Madras 600 035, India

First edition 1978

Second edition 1993

© 1978, 1993 B.A. Richardson

Typeset in $10^1/_2/12^1/_2$ pt Sabon
by Graphicraft Typesetters Ltd, Hong Kong
Printed in Great Britain at the University Press, Cambridge

ISBN 0 419 17490 7

A catalogue record for this book is available from the British Library

Library of Congress Cataloging-in-Publication data

Richardson, Barry A., 1937–
    Wood preservation / Barry A. Richardson. — 2nd ed.
        p. cm.
    Includes bibliographical references (p.) and index.
    ISBN 0–419–17490–7 (alk. paper)
    1. Wood—Preservation. I. Title.
TA422.R53 1993
674′.386—dc20   92–30664
                                                            CIP

To my friend John F. Levy whose
advice and encouragement have
been highly valued for many
years by all of us who are involved
in studies of wood deterioration and
preservation

Contents

Contents

Preface

Preparing a second edition of a technical book is always interesting because the alterations that are necessary indicate the amount of progress that has been made since the issue of the first edition. In this case I believe that very little progress has been made since I wrote my Preface to the first edition in December 1977, and certainly much less progress than was made in the previous 15 years, which were probably one of the most exciting periods in modern wood preservation development.

There are several reasons for this limited and disappointing progress in recent years. It is easy to forget that the world suddenly became aware that we were rapidly exhausting our reserves of hydrocarbon fuels, and shortages and escalating fuel prices then affected our lives and particularly our industrial operations. Whilst the use of petroleum solvents in preservatives was obviously discouraged, the shortage of energy affected the manufacture and application of all wood preservatives at a time when it was also being recognized that forest resources were being harvested faster than natural and plantation renewal, and when wood preservatives were therefore becoming particularly attractive as a means to reduce unnecessary deterioration and consumption. The energy crisis triggered an economic recession throughout the world and the wood preservation industry was affected in the same way as most other industries, and research and development expenditure has been drastically reduced. At the same time the development of new preservative systems has become increasingly difficult due to the introduction of more stringent health and environmental

controls. As a result only very large companies and consortia can now afford to develop new preservatives, and small companies are suffering serious difficulties as their established products become subject to increasing restriction or even prohibition.

This increasing awareness of health and environmental dangers has not necessarily resulted in the introduction of safer products, but instead the continuing use of products which have been widely accepted for many years. For example, widely used preservatives such as creosote and the copper–chromium–arsenic systems could not be introduced today, yet the development of safer alternatives is virtually impossible because of the enormous costs involved, even if a new preservative is based on established knowledge and experience. The present system is therefore actively discouraging the development of new preservatives which are more efficient and safer, and is instead encouraging, through economic necessity, the extended use of established and less safe systems.

I believe that, when we look back at this period in 15 or 20 years' time, we will consider it to be perhaps the most depressing period in the history of wood preservation, but I also hope that our health and environmental control systems will become more realistic, actively encouraging the development of safer systems but equally actively discouraging hazardous systems. One of the most serious problems with the present system is the lack of understanding of the hazards; controls are based so often on particular groups of toxic compounds without recognizing that some individual compounds are much less toxic than others, and without appreciating that the toxicity of a preservative does not depend on the presence of a toxic ingredient alone, but also on the concentration at which it is used. Perhaps I can help a little by including, in this completely revised second edition of this book, an entirely new section in which I discuss these health and environmental problems.

Barry A. Richardson
Latchmere
Lainston Close
Winchester
Hampshire SO22 5HJ

Preface to the first edition

Perhaps the most difficult task facing an author is to decide upon the type of person for whom he is writing. This book is an attempt to provide a reasonably comprehensive and non-controversial account of wood preservation of value to a person approaching a study of this subject for the first time, yet it is likely to be of equal value for reference purposes to the person who is already involved in commercial wood preservation or related research. Indeed, those in industry would naturally tend to specialize, perhaps concentrating upon certain preservation processes involving the use of a particular type of preservative applied by a particular method. Those involved in research are likely to be concerned with a limited geographical area and their interest will normally be confined within their own scientific discipline, such as entomology or mycology. To all these persons this book attempts to provide information on the other areas of wood preservation beyond their daily experience.

Wood Preservation is primarily an account of the situation in the principal temperate areas of Europe, North America, South Africa and Australasia, but the text refers in many respects initially to the situation in the British Isles, where wood preservation is most advanced. Wood has been imported into the British Isles for several centuries, so that it is widely accepted that it is valuable and preservation has long been economically justified. Wood preservation was first introduced as an industrial process in England and it has continued to be used in situations where decay is otherwise inevitable, such as for railway sleepers (ties) and transmission poles. However, it is not sufficient to confine this account to the British Isles alone, for even British readers require information on many other areas. As modern trade has expanded preservatives and preserved wood products have been exported to an increasing extent to countries with substantially different decay hazards. In addition new borers and fungi have been introduced on imported materials.

Wood Preservation is a book on a science (or is it an art or technology?) that is steadily developing, a fact that may be overlooked by scientists using this book, who will almost certainly criticize the lack of a bibliography and references to specific statements in the text. This cannot be accepted as a serious criticism as anyone with such an advanced interest in wood preservation will already be aware of the papers published in, for example, the Records of the Annual Conventions of the British Wood Preserving Association and the Proceedings of the American Wood Preservers' Association, which have extensive bibliographies and which provide a far more up-to-date source of further information than can be provided in any book. Other readers may criticize the failure to quote specifications for test methods, preservative formulations and treatment requirements, but again these are continuously revised and vary in each country so that the function of this book is simply to establish the principles involved, leaving the individual reader to obtain copies of appropriate specifications when required direct from the issuing authorities, such as the British Standards Institution, the Nordic Wood Preservation Council and the American Society for Testing and Materials. No doubt a further criticism will be the failure to comprehensively list proprietary preservatives but again these are subject to frequent changes; some are mentioned by name and a few are described in detail when it is considered that they are particularly important, but the enormous space required to list and describe the several thousand preservatives that are now available cannot be justified.

Wood Preservation is concerned with wood preservation, not with wood deterioration. Whilst it is obviously necessary for anyone involved in preservation to possess at least a basic knowledge of the deteriorating agencies that require to be controlled, the identification of deterioration is of limited importance. A reasonably detailed account of deteriorating organisms is given in the appendices, but these lack the diagnostic tables that are so often a feature of such descriptions; the identification of deterioration is the speciality of those who inspect structures and prepare specifications for remedial treatment, a subject that is considered in far greater detail in *Remedial Treatments in Buildings*. Although a section on wood structure is included in Chapter 1, it is assumed throughout that the reader has a basic knowledge of the properties and uses of woods. If this creates a

difficulty for any reader, he can refer to *Wood in Construction* or to Appendix A, which not only summarizes the most important preservation treatments but also includes a summary of the properties of the more important structural woods.

I was introduced to wood preservation by my father, Stanley A Richardson, and it is a subject that has always proved interesting to me. The more I know about wood preservation the more I become aware of our lack of knowledge and the need for further observations and investigations. Wood preservation is a remarkably complex subject, involving so many different disciplines, and my first impression upon completing the writing of this book was of the enormous amount of information that it had been necessary to omit and thus the very limited amount that could be included. I have always been encouraged in my studies of wood preservation by my many friends in industry and the related academic and research institutions, many of whom have kindly provided the illustrations that I have used. There are too many of them to list, but I would like to mention Dr John F. Levy of the Imperial College of Science and Technology, London, whose thoughts on deterioration and preservation are a stimulation to so many of us. We must give him credit for the very profound statement that 'as far as a fungus is concerned, wood consists of a large number of conveniently orientated holes surrounded by food', surely the most impressive statement ever in support of the need for scientific wood preservation.

Barry A. Richardson

Preservation technology

1.1 Introduction

It must be accepted that wood decay is inevitable. Indeed, if this were not the case our forests would soon become cluttered with the giant skeletons of dead trees. Natural durability is simply an indication of the rate of decay, but there is a further factor of fundamental importance – whilst decay may be inevitable in the forest it is not necessarily inevitable in wood in service. For example, fungal decay is dependent on an adequate moisture content, so that a structure designed to maintain wood in a dry condition is sufficient to ensure freedom from fungal decay, whatever the species of wood. In areas where wood borers exist which are capable of destroying dry wood, these structural precautions are insufficient and it becomes essential to select wood species which, whilst they may be susceptible to ultimate destruction from fungal decay, possess good natural resistance to the wood-borer concerned. If wood with adequate natural durability cannot be obtained it becomes necessary to adopt preservation processes, although these cannot be applied universally but only to those woods which are sufficiently permeable to permit the required penetration and retention of preservative.

Need for preservation

Preservation involves additional cost and must clearly be justified. The environmentalist may see preservation as a means for reducing our demand for replacement wood, thus conserving our forests. The economist may wish to conserve our forests for rather different reasons but the principle remains the same. Indeed, wood-importing countries will wish to preserve in order to conserve foreign currency by reducing wood imports, whilst wood-exporting countries will adopt preservation in order to reduce home demand for replacement wood, thus leaving the maximum possible volume available for export. Even in the most primitive tropical jungle village wood preservation has economic importance, for in these conditions the ravages of fungi, termites and other wood-destroying organisms ensure that an unacceptable amount of time and effort is devoted to replacing wooden structures such as homes and bridges. If preservation is practised, either by the selection of more durable species or by the adoption of a simple preservation process, structures may double their life. In this way more time and effort is available to improve the quality of life in the community, perhaps by growing extra crops for sale. In such primitive communities wood has no value as it is freely available, but the labour for repair and reconstruction represents a substantial burden on the community which is just as significant in more sophisticated countries. For example, in temperate climates a normal transmission pole pressure-treated with creosote will have a typical life of 45–60 years, whereas an identical untreated pole will last only 6–12 years. A similarly treated railway sleeper (tie) can be expected to last more than 35 years in comparison with only 8–10 years for untreated wood. In these conditions preservation has now been universally adopted and, as a result, there is a tendency to forget the basic economics; if untreated structural wood deteriorates the expense incurred is not confined to the cost of its replacement, or even this and the additional cost of labour required, but it also involves the perhaps much higher cost arising through structural failure. It can always be argued that failure can be avoided through regular inspection, but this cannot reduce the amount of disruption caused whilst services are interrupted during repair and replacement.

Preserved wood must be regarded as an entirely new structural material and must not be considered as just an improved form of wood, as it can be used in entirely different circumstances and certainly in more severe exposure situations. The most obvious advantage of preserved wood is that it can be used with impunity in situations where normal untreated species would inevitably decay, but it may be argued that, in many situations, this is a property that it enjoys together with many competitive materials. In fact, the use of wood has many advantages. It is extremely simple to fabricate structures from wood and, even in the most sophisticated production processes, the tooling costs are relatively low compared with those for competitive materials. Wood is ideal if it is necessary to erect an individual structure for a particular purpose but it is equally suitable for small batch or mass production. When these working properties are combined with the other advantages of wood, such as high strength to weight ratio, its excellent thermal insulation and fire resistance, and the unique aesthetic properties of finished wood, it sometimes becomes difficult to understand why alternative materials have ever been considered! However, there is one feature of wood which is unique amongst all structural materials; it is a crop which can be farmed, whereas its competitors such as stone, brick, metal and plastic are all derived from exhaustible mineral sources.

With all these various advantages wood has little to fear from competitive materials, provided it is efficiently utilized and either selected or preserved to ensure that it is completely durable in service. The need for durability is obvious, yet traditions are difficult to displace and in many countries the progressive deterioration of wood in service is generally accepted. It is unlikely that all owners of buildings and other wooden structures throughout the world can be educated to appreciate the actual costs of the material and labour involved in repairing decay damage, but the authorities in many countries are becoming increasingly conscious of the way in which these costs can affect prosperity. In this connection one current problem is the demand for wood pulp which directly competes with structural wood for the available forest resources. A high pulp yield can be achieved after short growing periods so that there is a tendency to fell forests whilst they are very immature to give rapid return on the invested capital. This has resulted in rapid increases in the cost of wood and a further justification for its efficient utilization and its preservation to avoid decay.

History of preservation

Preservation is not, in fact, new. The ancients worried little at first about decay as their buildings were seldom very permanent and replacement wood was easily obtained. Probably the person earliest recorded as using wood preservative was Noah who, when building the ark, was instructed by God to 'pitch it within and without with pitch'. In fact, various oils, tars and pitches were used from times of the most remote antiquity. Herodotus (c. 484–424 BC), a Greek whose monumental work earned him the title of 'Father of History', writes of the art of extracting oils, tars and resins. He also draws attention to a much older system of preserving organic matter, the ancient Egyptian art of mummifying or embalming bodies. This is probably the most efficient method of preserving organic matter that has ever been devised. The Egyptian mummies are now at least 4000 years old and many are as well preserved as when originally entombed. Herodotus and Diodorus Siculus (1st century BC) indicate that the body was steeped in natrum (or natron) for 70 days and then in an oily or bituminous substance for a similar time. Natrum, the production and use of which was a state monopoly in Ptolemaic times, from *c.* 320 BC, was a mixed solution of sodium sesquicarbonate, chloride and sulphate. It was obtained from three centres fed during the flood season by seepage from the River Nile. The most important centre was an oasis in the Western Desert still known as Wadi Natrum. It is not possible that mummifying was practised in the simple way described as the bituminous substance would scarcely penetrate, yet it has been found that even the interior of the bones has been penetrated. It is probable that the body, after steeping in natrum, was placed in the bituminous substance which was heated to temperature above the boiling point of water so that the water within the body volatilized and was then replaced by the oil. Boulton carried this out on a piece of wood in the middle of the 19th century; his results indicated the correctness of the theory and were also the origin of the Boultonizing treatment which is still in use today.

The Egyptians were not the only people to use metallic salts as preservatives. The Chinese were immersing wood in sea water or the water of salt lakes prior to use as a building material before 100 BC. Well preserved props have been removed from old Roman mines in Cyprus and examination has shown them to contain metallic copper, well distributed throughout both the heart and sapwood. Various theories have been advanced to explain its presence as a Roman attempt at preservation, but is seems more likely that the true explanation involves the copper found in the soil in this area. It is possible that the process was electrolytic, one end of the prop being in one type of soil containing copper and the other end being in another type of soil so that the damp wood formed a rather complex cell.

Marcus Porcius Cato (234–149 BC), a Roman whose condemnation of the luxury of his times earned him the nickname of 'Cato Censorious', commented on wood preservation, but by far the most informative writer was the renowned Roman naturalist Pliny. Pliny the Elder (AD 23–79), who perished at Pompeii during the eruption of Vesuvius, mentioned that Amurca, the oil-less by-product in the manufacture of olive oil, and also oils of cedar, larch, juniper and nard-bush (*Valeriana* spp.) were used to preserve articles of value from decay. He claimed that wood well rubbed with oil of cedar was proof against woodworm and decay, and in his writings described the preparation of 48 different kinds of oil for wood preserving. He also observed that the more odoriferous or resinous the wood the more resistant it was to decay. Because the statue of Zeus (Jupiter) by Phidias was erected in a damp grove at Olympia its wooden platform was imbued with oil. The statue of Diana at Ephesus was made of wood and was believed to have been of miraculous origin. Pliny, quoting an eye-witness Musicians, notes that it was still thought necessary to saturate it with oil of nard through small orifices bored in the woodwork. The Roman use of olive oil was copied from Alexander the Great (356–323 BC), the king of Macedon who conquered a large area of the known world. He is said to have ordered piles and other bridge timbers to be covered with olive oil as a precaution against decay.

The previously mentioned statue of Diana at Ephesus was underpinned with charred piles. This was not a new idea as a prehistoric race, the Beakermen, applied charring to wood. The aborigines called the Tiwi who live on Melville Island near Darwin, Australia, and whose civilization is said to be 50 000 years behind our own, mark their graves with brightly painted poles like American Indian totem poles, which are made from Bloodwood, a hard red-sapped wood which they have found to be resistant to termites and fungal decay. Prior to painting, the wood is charred and covered with beeswax, orchid sap or white of turtle egg. This may be the continuation of some primitive knowledge of wood preservation. Alternatively it may be solely a method of forming a suitable background for painting.

As soon as man began using wood as a building material it was only a matter of time before decay became domesticated. The fungus that probably causes the greatest damage in buildings is the common Dry rot fungus *Serpula lacrymans*. It has never been recorded as occurring in nature and appears to be associated only with man-made structures. The name is derived from the Latin word *lacrima*, a tear, for *Serpula lacrymans*, (formerly known as *Merulius lacrymans* is the weeping fungus as fresh growth can often be observed covered with drops of water. It is this weeping or fretting which enables it to be identified as the 'fretting leprosy of the house' in the Old Testament Book of Leviticus. Until comparatively recent times the priest was the person who was called in to deal with any kind of trouble or pestilence and consequently Leviticus contains full instructions on how to deal with the 'leprosy of the house'. The priest, when carrying out his inspection, was to look for 'hollow strakes, greenish or reddish, on the walls'. If they were present the house was to be shut up for 7 days, and if after that time 'the plague be spread in the house, it is a fretting leprosy' and Dry rot rather than one of the Wet rots, and he was to 'command that they take away the stones in which the plague is, and . . . cast them into an unclean place without the city, . . . the house to be scraped within round about, and . . . pour out the dust . . . without the city into an unclean place'. This may appear rather ruthless but even today an affected area

must be stripped to the bare masonry to ensure the successful application of fungicide, the Dry rot fungus spreading through masonry in the search for wood. It is in these verses of Leviticus, which also describe leprosy in man, that we may read of early ideas of contagion, and people entering the house were required to wash themselves and their garments thoroughly on leaving. If the priest found that the fungus had not developed after replastering the affected area he was to apply final 'fungicidal' treatment and 'take to cleanse the house two birds, and cedar wood, and scarlet, and hyssop'. He was instructed to sacrifice one bird and sprinkle the house seven times with the blood. The other bird, after being dipped in the blood, was freed and flew away, presumably taking the pestilence or infection with it!

It was the belief that the words of the Bible and several other books were completely irreproachable that severely discouraged the development of science and technology up to the early 16th century. Ideas not in agreement with these standard works were considered heretical, and often people were put to death for expressing them. Other than those already quoted, references to timber preservation before the 18th century appear to be negligible, although timber decay was frequently described and appears to have been a serious problem.

Early problems

In the reign of Elizabeth I, Britain's greatest asset was her navy. When Elizabeth came to the throne she found that ten out of the 32 Royal ships were suffering from decay. There was, apparently, no accepted method of wood preservation, and the condition of the navy was so bad a few years later during the reign of James I that a commission of inquiry was appointed. The findings emphasized the importance of constructing ships of seasoned wood, but little notice appears to have been taken of this. James I was succeeded by his son Charles I, who was beheaded, but whose son Charles

II, was crowned on the restoration of the monarchy. A period of rearmament commenced, the navy's programme of shipbuilding being in the control of Samuel Pepys, the Secretary to the Admiralty. In his famous diary Pepys remarked on the shortage of wood being such that a large amount of green and unseasoned wood was used in shipbuilding. The result was that many ships began to decay before being commissioned. Pepys very wisely suggested that this would not be so if the ships were better ventilated. These troubles were not confined to Royal Navy ships, and the merchantmen of the East India Company seldom made more than four, and sometimes only three, voyages to India before becoming useless through decay.

The shortage of wood that Pepys referred to was becoming serious. The almost continuous arming and rearming occurring throughout these times meant that extremely large quantities of wood were required for shipbuilding. Concurrently the rise in the use of iron saw the progressive destruction of the Wealden and other forests to provide smelting charcoal. Daniel Defoe in 1724 wrote that the Sussex ironworks were carried on 'at such a prodigious expense of wood, that even in a country almost overrun with timber, they began to complain of consuming of it for the furnaces and leaving the next age to want for timber for building'. He saw no justification for the complaint, for Kent, Sussex and Hampshire were 'one inexhaustible storehouse of timber'. Defoe was wrong, of course, for wood was being used at such an alarming rate that there was already a noticeable shortage in the 16th century. Various Acts were passed through Parliament in attempts to limit consumption and obtain supplies from abroad. There were also attempts to transfer the iron smelting industry to North America but this never came about because of the introduction of coal for smelting. Later a further significant economy occurred when prejudices against the use of coal as a domestic fuel were finally overcome. Softwood began to arrive in

increasing quantities from the Baltic and Canada but, despite this, the general shortage soon raised prices and the need for preservation became more apparent.

The wastage of ships in the Royal Navy on account of decay was becoming an extremely grave problem. Little seems to have been done against Dry rot, for in 1771 Lord Sandwich had the fungus dug out of the reserve ships so that he might inspect the timbers. Dry rot was not the only problem the navy had to contend with, for there were also the marine-borers such as shipworm and other fouling organisms that attacked the outsides of ships. At the time of Vasco de Gama (1469–1524) the Portuguese are known to have charred the outsides of their ships as a protection against shipworm, and in 1720 the Royal Navy built a ship the *Royal William* entirely from charred wood. It appears that the experiment failed for it was never repeated, but charred wood is still used for shipbuilding by the Solomon Islanders, who apparently learned the practice from the Portuguese. Covering the hull with sheet metal as a protection against shipworm was a system used as early as Roman times. Lead was the metal that was most easily worked into sheets but it was so heavy that it pulled away from its fixings. Copper was tried by the British Navy but this coincided with the introduction on a large scale of iron fittings for ships. Electrolytic action occurred, seriously damaging rudder bearings, so that the use of metal sheeting was discontinued. In 1782 the prevalence of Dry rot in Royal Navy ships was made very apparent by the tragic loss of the *Royal George* at Portsmouth. At the court martial investigating the deaths of the 800 men on board it was disclosed that previously the bottom had fallen out while the ship was being heeled over for slight repairs.

Early preservation

In 1784 the Royal Society of Arts offered a gold medal 'for the discovery of the various causes of Dry rot in timber and the certain method of its prevention'. It was awarded in 1794 to Batson who, in 1778, had treated an outbreak of Dry rot in a house by removing sub-floor soil and replacing it with anchor-smith's ashes. In the early 19th century Britain was successfully negotiating a most difficult period of history, highlighted by the wars against Napoleon and in North America. The Royal Navy was more important than ever before but, although the shortage of available wood brought the decay of ships daily into clearer perspective, it was not until 1821 that the Admiralty asked James Sowerby to investigate the problem. He reported on the state of the *Queen Charlotte*, a first rater of 110 guns that was launched in 1810 at a cost of £88 534. Only 14 months after launching she had to be re-built at a cost of £94 499 before she could be commissioned. By 1859, when her name was changed to *Excellent*, the total cost of repairs had risen to £287 837. Sowerby reported the cause as fungal rot and identified a score or more of fungi.

The Royal Navy was not the only sufferer from Dry rot. In 1807 James Randell spoke to the Royal Society of Arts on the subject of Dry rot and mentioned that it had destroyed the great dome that Robert Taylor had built on the Bank of England. By this time people were beginning to take serious interest in chemical preservatives. 1812 saw the first attempt at fumigation when Lukin experimented in Woolwich dockyard with the injection of resinous vapours; the attempt was abandoned after the apparatus exploded with fatal consequences to the workmen.

Salt preservatives

The 19th century produced further interest in wood decay on account of the widespread expansion of the railways. The stone blocks first used for supporting the rails were found to be too rigid and wooden sleepers were substituted. These rapidly decayed and obviously a good chemical preservative was required. The first

list of established preservatives, published in 1770 by Sir John Pringle, was followed shortly afterwards by a second list drawn up by Dr Macbride. The age of chemical wood preservation had arrived and these lists both appeared in the 1810 edition of *Encyclopaedia Britannica*. By 1842 five processes were established using mercuric chloride, copper sulphate, zinc chloride, ferrous sulphate with a sulphide, and creosote respectively.

Mercuric chloride, first used by the French scientist Homberg in 1705 to preserve wood from insect attack, was later recommended by De Boissieu (1767) and Sir Humphrey Davy (1824), and in 1832 Kyan took out his patent for this process which became known as Kyanizing. Kyan's first success was the preservation of the Duke of Devonshire's conservatories. The British Admiralty tested it and found it failed against marine organisms. This was more than a century after the Dutch Government had found exactly the same result. Mercuric chloride, also known as corrosive sublimate, is soluble in water, volatile at ordinary temperatures and poisonous. Its use continued for some time in the United States and Germany but probably the last large-scale instance of Kyanizing in Britain was in 1863.

The use of copper sulphate was recommended as early as 1767 by De Boissieu and Bordenave. Thomas Wade recommended it in 1815, and in 1837 Margary took out a patent for its use in wood preservation. Copper sulphate was by far the most successful metallic salt and long after it died out of use in England its popularity continued in France, where it was applied by a most ingenious method known as the Boucherie process, by which a newly felled tree is impregnated by replacement of the sap. The Boucherie copper sulphate process was very popular in France from about 1860 until it went out of favour in 1910 because of some failures on alkaline soil. It was used for preserving telegraph poles in Britain prior to nationalization in 1870, and continued in use in the South of

France and Switzerland until a few years ago. The Boucherie process was revived in 1935 by the Deutsche Reichpost but with a multisalt preservative.

The antiseptic properties of copper sulphate have never been questioned and it is only the solubility in water of this and other metallic salts that makes them unsuitable as wood preservatives in wet situations. It will be recalled that the ancient Egyptians had a very effective method for the preservation of bodies, which Boulton suggested had involved steeping in natron followed by pickling in a bituminous substance. Boulton impregnated a piece of wood with natron and afterwards placed it in creosote at a temperature above the boiling point of water. He found that water evaporated, depositing the natron salts in the wood, and the creosote then penetrated well into the dried wood. This process formed the basis of a patent by Boulton in 1879 except that copper sulphate was used in place of natron. It was successful but was little used because of the success of creosote alone, the addition of the copper sulphate serving only to increase the cost. Boulton suggested, however, that an oil with no preservative properties could be used in place of the creosote, its purpose being solely to prevent leaching of the copper sulphate. Modern Boultonizing involves the use of high-temperature creosote and vacuum, but simply to boil off moisture within the wood so that the creosote is able to penetrate.

Zinc chloride was recommended as a wood preservative in 1815 and 1837 by Thomas Wade and Boucherie respectively. In 1838 Sir William Burnett patented its use but throughout its wood preserving history it has suffered because of its extreme solubility in water. Despite this it was much used by the Royal Navy. Because of the shortage of creosote in the United States its use continued there long after it was forgotten in Britain. Even there, however, its use gradually diminished because of the expansion of the coal gas industry and the increased imports of

creosote from Britain as return cargo in petroleum tankers. Although extremely soluble it was found to retain its power of preservation to a small extent. This was thought to be due to the formation of zinc oxychloride, insoluble in water but poisonous to fungi because of its solubility in their enzyme secretions. This hypothesis suggested the idea of the formation of insoluble preservatives within wood by the application of two or more solutions, a process that has been effectively applied using various materials since the beginning of the 20th century. However, the first precipitation process was a complete failure. It was in 1841 that Payne was granted a patent for his two-stage process which involved impregnation of wood with ferrous sulphate followed by calcium sulphide. It was said that a double decomposition occurred within the pores of the wood, forming ferrous sulphide and calcium sulphate, both only sparingly soluble in water, but the treatment was found to have negligible preservative effect.

Coal-tar products

Creosote is certainly the most successful of the preservatives developed during the 19th century. The process, known as creosoting, is based essentially on a patent granted to John Bethell in 1838. Bethell's patent lists 18 substances, mixtures or solutions, including metallic salts, oleaginous and bituminous substances. Although the word 'creosote' is not used, mention is made of a mixture of dead oil with two or three parts of coal-tar, and this is the origin of creosoting by pressure impregnation. It was not the first attempt at the use of tars, for as early as 1756 experiments were carried out in Great Britain and America (Knowles) on the impregnation of wood with vegetable tars and extracts. Boulton suggested that the term creosote originated in a patent granted to Franz Moll in 1836. This patent was concerned with injecting wood in closed iron vessels with extracts of coal-tar, first as vapour and then as oil in a liquid state. Moll termed oils lighter than water 'Eupion' and those heavier 'Kreosot', the latter being said to have antiseptic qualities. The process was not practical as the light oils immediately evaporated on the application of the heated kreosot, and it was left to Bethell to patent the modern creosoting process two years later. Franz Moll probably derived his term 'kreosot' from the Greek words *kreas* for flesh and *soter* for preserver but, although the term kreosot was applied to the heavy coal-tar oils, the term 'creosote' was not derived directly from it. Even in Franz Moll's time the term 'creosote' was applied to a product of the dry distillation of wood, and the term was applied to the heavy tar-oils in the belief that true creosote was identical to the carbolic acid contained in coaltar. Boulton mentions that Tidy compared these two substances and showed them to be dissimilar. He also demonstrated that coal-tar contained no true creosote but the term 'creosote' is now universally applied to the heavy oils produced during the distillation of coal-tar.

By 1853 creosote had established itself as a most reliable and persistent wood preservative, and most other processes were abandoned. In France, however, creosote established later because of the popularity of copper sulphate allied by the Boucherie process. In 1867 Forestier, working for the French Government and also Dutch Government investigators, showed that creosote of a suitable grade, efficiently applied, rendered wood resistant to shipworm. At about the same time, Crepin, working for the Belgian Government, showed that this applied also to other marine animals. However, there were distinct failures resulting from the use of the wrong type of creosote and these focused attention on studies of the composition and effectiveness of various grades.

In 1834 the German chemist Runge discovered phenol (carbolic acid) in coal-tar, and in 1860 Letheby attributed the preservative properties of coal-tars to this component. Carbolic acid was recognized as an effective fungicide and it was due to investigations into the wood-preservative properties of coal-tar that the world gained the benefit of a most important medical development. A young surgeon, Joseph Lister, was discussing with a railway engineer his method for preserving sleepers (ties) and was informed that it was the carbolic acid which prevented decay. Lister, worried by the high death rate due to infection during operations, immediately saw that carbolic acid might be used to prevent it. He started operating under a continuous carbolic acid spray, all his instruments and his own hands having been washed in the acid. Conditions were most unpleasant but he thought it worthwhile if he could save a few lives. He was surprised and delighted to find that, instead of the slight improvement that he had hoped for, he had achieved almost complete success. There was opposition to his idea at first, but his dramatic results eventually gained him universal recognition and it was not long before carbolic acid was described in medical textbooks as the 'aerial disinfectant par excellence'.

Letheby, appreciating the antiseptic properties of carbolic acid, specified that naphthalene and para-naphthalene should be excluded as far as possible from creosote as he considered them to have no preservative value. However, in 1862 Rottier concluded that carbolic acid, although an energetic antiseptic, had little persistent effect due to its volatility and solubility in water. He attributed the durable success of creosote to the heavier and less volatile components of coaltar. Despite the interest in wood preservation resulting from the expansion of the railways and later of the telegraphs, it was not until 1863 that really instructive experiments were carried out. These experiments, which were conducted by Coisne on behalf of the Belgian Government, were repeated in 1866. Coisne treated wood shavings with various grades of creosote and placed them in a putrefying pit for 4 years. The results were entirely in favour of the use of the heavier oils, tar acids by themselves having no persistent effect. These results were confirmed by long-term experience and the Belgian Government adopted the recommendations for its successful creosoting specifications.

About this time there were two theories on the cause of decay. The accepted theory was that of the great scientist Liebig, enunciated in great detail in 1847 and 1851, which claimed that putrefaction or decay of an organic material was a form of slow combustion which he termed 'Eramacausis' and that it was initiated on contact with bodies which were already affected. He discovered that it could be prevented by lack of moisture and atmospheric air, and from this he deduced (and later showed) that it was provoked by oxygen. Dalton's atomic theory, proposed in 1808–1810, was by this time one of the foundation stones of science and Liebig claimed that the method of transfer was the communication of motion from atoms of the infected matter to atoms in the contacting material. He denied that fermentation, putrefaction or decomposition was caused by any fungi, animalcules, parasites or infusoria that might be present, their presence being coincidental or due to a preference for feeding on decaying matter.

The parts of animals and plants which decay most rapidly are the blood and the sap. It was suggested that decay could be prevented by coagulants of albumin, such as mercuric chloride, copper sulphate, zinc chloride and the tar-oils. In 1854 Louis Pasteur was appointed Professor and Dean of the Faculty of Science at Lille. Here he concentrated on his study of fermentation in the production of beer and wine. Three years later he moved to the Ecole Normale at Paris as Director of Scientific Studies, and while there he proclaimed that fermentation was the result of the action of minute organisms. If fermentation failed to occur it meant that the organism was absent or unable to develop properly.

Liebig had observed that decay required atmospheric air and deduced that this was because oxygen was necessary. He confirmed this theory by showing that oxygen accelerated decay. Pasteur repeated the experiments and in 1864 announced that decay was caused by minute organisms that were not spontaneously generated but were instead present in atmosphere, and this was one reason why atmospheric air was necessary. It was only a year after this that Lister appreciated the significance of infection in surgery and, as previously mentioned, initiated the use of carbolic acid as a surgical disinfectant. Pasteur's theory was confirmed by Koch and soon gained support among authorities such as Tyndall. This was immediately thought to be a simplification of the theory of preservation as the only problem was to discover substances toxic to the decay-causing organisms.

Pasteur was not the first to declare decay to be caused by living organisms. As far as wood was concerned there was the obvious damage by borers and also the presence of fungi. In 1803 Benjamin Jonson had declared Dry rot to be the result of a 'visit from a plant and is and ever was so' but he left it to Theodore Hartig in 1833 to recognize the general association between fungi and decay. This association had been noticed before but it was Theodore's son Robert who in 1878 showed fungi to be directly responsible for decay. He continued his studies in 1885 and made a thorough investigation into the Dry rot fungus and its effect on wood.

By 1884 the wood-preserving industry had been established long enough for serious interest to be taken in long-term effectiveness. The metallic salts had broken down completely and their use had been largely discontinued in favour of creosote. It was many years before interest in salt preservatives revived with the development of the multisalt preservatives, described in detail in Chapter 4. Creosote had been known to fail but because of the careful records of treatment kept by some of the impregnating companies, coupled with the work of people like Coisne, Boulton and Tidy, specifications had been developed which could be relied on to give good protection. Boulton carried out tests in 1884 on a 29-year-old fence in London Docks, apparently as sound as when it was erected. He detected no tar acids but found the semi-solid constituents of tar-oils, including naphthalene, to be present. He found very little distilling below 232°C (450°F) and 60–70% distilled above 316°C (600°F). He managed to detect acridine solidified in the pores of some of the specimens. This is an acrid and pungent substance, neither volatile nor soluble in water, that had been discovered by Graebe and Caro. Greville Williams also examined samples from the fence and, although he managed to detect traces of tar-acids, the indication was very slight and was probably due to the heaviest tar-acids trapped within solidified portions of the oil. In nearly all of the specimens he detected naphthalene and in all he detected acridine and basic substances. He concluded that the preservative action was due more to the latter than to the tar-acids. Tidy experimented on naphthalene, finding that it remained in the pores of the wood. Although not as powerful an antiseptic as the tar-acids it was far more persistent. He decided that the para-naphthalene or anthracene contained in tar-oils was probably without wood-preserving properties and drew up his creosote specification accordingly. This standard, introduced in 1883, was the basis of nearly all British specifications until the BESA (now BSI) specification was introduced in 1921.

In 1824 Hennell had synthesized alcohol and 2 years later Wöhler was responsible for the synthesis of urea. These achievements opened the door to tremendous developments in industrial synthesis of organic compounds. Coal is a veritable treasure chest of raw materials for these processes, and it was not long before coal-tar began to suffer from the extraction of some of its components. Typical of this was the use made of anthracene. From the earliest times the roots of madder (*Rubia tinctoria*) had been used as a dyestuff in India and Egypt. The principal dye involved is alizarin which is present in the root as a glucoside, ruberythric acid. This can be hydrolysed to glucose and alizarin, and was extensively employed until towards the end of the 19th century in the production of Turkey Red dye for dyers and printers. However, in 1868 Graebe and Lieberman found that alizarin could be reduced to anthracene by heating with zinc dust. They suggested a rather expensive process for synthesizing alizarin from anthracene, which was soon relinquished in favour of an alternative process they discovered simultaneously with Perkin, the 'Father of Dyeing'.

Whilst the increasing sophistication of the chemical industry threatened to reduce the effectiveness of creosote, it was also ultimately responsible for the development of compounds such as pentachlorophenol and the organo-chlorine insecticides which made the formulation of organic solvent-based preservatives possible, as described in Chapter 4. Fortunately, Tidy had already shown that anthracene had only weak wood-preserving properties, so that there was no conflict between dye manufacturers and creosote users. Other changes in the composition of creosote were caused by the different methods of coking and the varying grades of coal. All this made it more important that the principal wood-preserving components in creosote should be identified. Work has continued to the present day, but despite improved methods the preservative action of creosote is still imperfectly understood. In 1951 Mayfield concluded that 'the toxicity of creosote is not due to one or a very few highly effective materials but is due to the many and varied compounds which occur throughout the boiling range. The value of creosote as a wood preservative depends largely on whether or not it remains in the wood under the conditions and throughout the period of service'. Essentially this means that a particular grade of creosote cannot be said to be efficient on the merits of its chemical composition alone. The only true test is to use it and see how it performs in normal service, but the difficulty is the length of life expected of creosote; the fence tested by Boulton in 1884 lasted about 70 years. Even then it was demolished only to make way for another structure and was still reasonably preserved. Any field test would take as long, so that evaluation of new preservatives is often based on laboratory comparisons of preservative toxicity.

Application methods

Little has been said of the methods used for applying preservatives. An effective preservative can be a complete failure if inefficiently applied, and this is the explanation of the early failures of creosote in the United States. Vacuum and pressure methods of impregnation undoubtedly give the greatest certainty of lasting preservation. Breant is said to have been the inventor of this process when he took out a patent in 1831, but in Great Britain Bethell was granted a patent in 1838 which included amongst other substances creosote applied by this means. The method soon became known as the full-cell or Bethell process, although it was modified to its present commercial form, which will be described in detail in Chapter 3, by Burt, who was granted a patent for his improvements to the method. With creosote the method is ineffective when applied to unseasoned or wet wood, so that extensive storage facilities are required for drying and seasoning. In 1879 Boulton was granted a patent for his Boiling under Vacuum process, using hot creosote to boil off the water in the wood. This process may be followed by the full-cell process or an empty-cell process such as the Rüping process. Steaming and steaming-and-vacuum processes were tried as alternatives to the Boulton process but with no great success.

There are several difficulties encountered with the full-cell process. Creosote bleeding is likely to occur, an annoying factor with fences and poles that pedestrians and animals are likely to encounter. Another aspect is the quantity of preservative used, a very important point in countries where preservatives, especially creosote, are scarce and expensive. The empty-cell processes are a great improvement as bleeding is less likely to occur and there is a 40–60% reduction in the use of preservative. The latter is especially important in the case of particularly permeable woods and those with a high proportion of sapwood. The empty-cell methods in common use, the Rüping and Lowry processes, will be decribed in Chapter 3.

The Rüping process was initially patented by Wasserman in Germany in 1902, although Rüping applied the process commercially and American patents were subsequently granted in his name. The process is commenced by the application of an initial air pressure. When the entire process is complete the pressure is released, the compressed air in the cells drives out some of the preservative and a short period of vacuum recovers more preservative, so that the net retention in the wood is only about 40% of the gross absorption, a saving in preservative of 60%. The Lowry process, which was patented in America in 1906, differs only in that it relies on compression of air at atmospheric pressure for return of excess preservative, so that there is no initial compression stage. The recovery of preservative is about 40%.

Other similar processes due to Hülsbert, including the Nordheim process of 1907, have been entirely superseded by the Rüping process. In 1912 Rütgerswerke AG were granted a patent for treatment of insufficiently dry timber by the Rüping process. It is identical with Boulton's patent except that an oil, used for evaporating the water, is drawn off before the Rüping process is applied. The vacuum and pressure methods are the most important and most effective methods used for the application of wood preservatives. They suffer, however, from the great disadvantage that special plant is required and it is often impossible or uneconomical to send wood to the plant for treatment. Numerous non-pressure methods are available but are suitable for use only with specially developed preservatives such as the low-viscosity organic solvent products for spray and dip treatment of dry wood, and the concentrated borate solutions which can be used for diffusion treatment of freshly felled wood with high moisture content. Preservation processes are discussed in detail in Chapter 3.

1.2 Preservation principles

The simplest method to avoid deterioration is to use only naturally durable wood. Durability is an embarrassment in nature as it delays the disposal of dead trees, and it can therefore be appreciated that only a limited number of wood species are truly durable. This durability is always confined to the heartwood but the elimination of sapwood, coupled with selection from a very limited range of species, is unrealistic unless very high costs can be tolerated. It is far more realistic to select a wood species for its physical properties and then to take suitable precautions to ensure that deterioration is avoided. This does not necessarily mean the use of preservative treatments. For example, the most efficient method to avoid fungal decay is to keep wood dry, and this is most simply achieved by structural design, such as the incorporation of overhanging eaves and gutters to dispose of rainfall and damp-proof membranes to isolate structural wood from dampness in the soil or supporting structure. However, there are some situations, usually termed *severe hazard* conditions, where deterioration is unavoidable unless naturally durable or adequately preserved wood is used. The most important severe hazard risk is the ground contact condition which arises in transmission poles, fence posts and railway sleepers (ties). In some areas insect-borer attack is virtually inevitable whatever the service conditions, such as in areas subject to the Dry Wood termites. In some parts of Europe the House Longhorn beetle, sometimes known as the House borer, represents a

severe hazard to softwood. In other situations deterioration may not be inevitable, yet it may be possible or even probable, representing a *moderate hazard*. Thus the Common Furniture beetle is a particularly widespread cause of damage, yet it seldom results in structural collapse. Similarly, fungal decay may not normally present a risk, yet it may be able to develop if structural woodwork becomes wet through accident or neglect.

It is obviously important to identify the deterioration hazard before deciding on the precautions that are necessary. However, the hazard does not vary only with the conditions to which the wood is exposed but also with the wood species. For example, a group of Basidiomycetes are responsible for the fungal decay that is commonly known as Wet rot. The Cellar rot *Coniophore puteana* occurs in persistently damp conditions, such as when a damp-proof membrane is omitted and when plates under floor joists are in direct contact with damp supporting walls. If the moisture content tends to fluctuate, as in wood affected by a periodic roof leak, the White Pore fungus *Poria vaporaria* is far more common in softwoods and, for example, the Stringy Oak rot *Phellinus megaloporus* in oak. *Coriolus versicolor* sometimes develops when non-durable tropical hardwoods are used as drips or sills on window and door frames, and *Paxillus panuoides* generally occurs where the conditions are too wet for these other fungi. A knowledge of the basic nature of various wood species, and perhaps even of the principles for their identification, is therefore essential for a proper understanding of the decay hazard.

The reliability of preservation processes also varies widely with the wood species. The most important requirement is to achieve an adequate retention of the preservative within the wood. In many species this can be achieved relatively easily in the sapwood but the heartwood may be completely impermeable. In Baltic redwood (Scots pine) the treatment of the sapwood may be all that is necessary as the heartwood possesses significant natural durability. In other species such a Baltic whitewood (spruce) even the heartwood is non-durable, yet neither heartwood or sapwood is sufficiently permeable to permit adequate preservative penetration. Preservative efficacy also varies with the microscopic structure of the wood. Thus the usually reliable copper–chromium–arsenic waterborne preservatives are much less efficient in hardwoods than in softwoods, apparently through the inadequate micro-distribution of the preservative within the cell walls. Clearly a detailed knowledge of the fine structure of wood is necessary if these various problems are to be fully understood.

1.3 Wood structure

The tree

The basic structure of wood, the variation between softwoods and hardwoods, the differences between species and the significance of various features ar all described in greater detail in the book *Wood in Construction* by the present author. Many of the features are of importance in wood decay and preservation. Wood is the natural supporting skeleton of larger plants and it is important to understand the origin of the skeletal parts in order to fully appreciate their properties. A plant consists of a crown of leaves, a supporting stem and the roots that anchor it within the soil. A tree is special only in regard to the scale of its development, and thus the need for a supporting skeleton which ultimately becomes the wood of commerce. However, the trunk does not perform solely this passive supporting function but also acts as a storage area, and the outer zones are the conducting routes between the crown and the roots. In addition, the growth of the crown must be accompanied by similar growth in the roots, and an appropriate enlargement in the trunk to enable it to continue to perform its supporting function.

The growth arrangement of a tree comprising a crown of leaves connected to the roots by usually a single main stem, trunk or bowl is known as the dendroid habit. The sole purpose of this very elaborate structure is simply to survive and to supply the cells within the plants. This is achieved firstly by the roots, which absorb water containing dissolved mineral salts which is then conveyed by the trunk, branches and twigs to the crown and the individual leaves. The function of these leaves is to absorb atmospheric carbon dioxide, which is combined with the water from the roots to form simple sugars by the process known as photosynthesis; the chlorophyll in green plants is the essential catalyst which enables this process to proceed whenever adequate ultraviolet radiation is received from the sun. The sugars are then conveyed throughout the plant to the leaves, twigs, branches, trunk and roots. The primary function of the sugars is to provide an energy source or food for the individual living cells within all these components of the tree, but a secondary function is to provide the basic units from which the skeletal structure of the tree is formed. Whenever there is sufficient sunlight the simple sugars will be produced by the leaves and will be found distributed throughout the living tissue of the tree. Some of this sugar will be formed into starch and deposited within the living tissue as a reserve energy source which can be utilized by the cells whenever sugar is unavailable from the leaves, such as at night or during the winter months when deciduous trees shed their leaves. Finally the sugar units are joined into cellulose chains which are then assembled into the main skeleton of the woody parts of the tree.

Wood formation

A tree is continuously increasing in size (Fig. 1.1) and this is the function of the embryonic tissue distributed around the whole plant. The increase in the overall size of the crown is the result of the activity of the apical meristem or

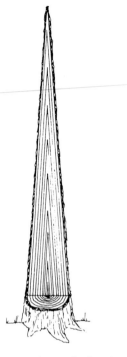

FIGURE 1.1 Diagrammatic trunk showing annual rings.

bud at the end of each twig which achieves the progressive extension in length. The detailed structure of this bud has little significance on the structure of wood but the meristem tissue actually extends over the entire surface of the tree, just beneath the bark of the twigs, branches and trunk but extending similarly over the entire root system. The purpose of this lateral meristem is to enable all the structural components of the tree to increase in girth so that they are capable of supporting the enlarged crown. Each twig as it lengthens consists initially only of pith formed by the apical meristem, but it is covered externally by the lateral meristem to permit subsequent increase in girth, although it also provides a protective covering to the new shoot to control water loss and to prevent disease damage. As this meristem tissue generates new cells which increase the girth the protective covering splits, exposing inner tissue, but the meristem tissue then generates a new protective layer which becomes the rough outer bark of the branches and trunk (Fig. 1.2).

A trunk in its simplest form could thus consist of single twig, progressively increasing in

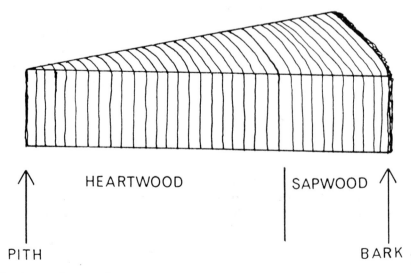

HEARTWOOD   SAPWOOD

PITH   BARK

FIGURE 1.2 Wood zones in a trunk.

length or height as it is extended by the bud at its apex, the lateral meristem also increasing the girth, so that the trunk possess a steep conical shape which is essentially the portion of the tree which is the wood of commerce. The trunk consists of a central pith enclosed, in effect, by a series of cones, each cone representing the annual growth increment or annual ring. The wood tissue around the pith is the heartwood and consists of dead cells. The heartwood is surrounded by the living cells and the sapwood or xylem which is covered by a thin layer of phloem cells and the protective bark. The interface between the xylem and phloem cells is known as the cambium, the term used by wood technologists for the actively dividing cells which are described by botanists as the lateral meristem. These dividing cambium cells are known as fusiform initials, the cells splitting off on the inner side of the cambium forming becoming xylem or wood tissue and those on the outer side forming phloem or bark tissue. In coniferous trees the xylem cells are termed tracheids whilst in dicotyledons or broad-leaf trees they are termed fibres. The cambium also possess additional active cells known as ray initials which generate horizontal radial bands of cells known as rays or parenchyma tissue within the xylem.

This description is not very complex, yet it explains the development of wood within the trunk of a tree. Xylem can be formed only when the entire tree structure is active. The xylem deposits thus tend to be seasonal, particularly in temperate areas, so that a trunk will increase each year by the addition of an outer cone of tissue representing the growth for a season. This growth varies from a wide band of large but thin-walled cells known as the early or spring growth, to much smaller thick-walled cells which represent the summer or autumn growth. This late growth terminates sharply as cell formation discontinues with the onset of winter and this sharp terminal line is then followed by the large

cells of the early growth for the following season.

These details explain the basic structure of wood but they have additional significance. The xylem or sapwood is the tissue that conducts water and dissolved salts from the roots to the crown of the tree, whilst the phloem is the tissue that conducts sugars from the crown to the various growing cells throughout the structure of the tree. When a xylem cell, a tracheid or fibre, is first formed it consists of a thin wall of sugars which have polymerized into cellulose. The sugars from the phloem continue to be supplied to xylem cells, the rays perhaps providing routes for this transfer, so that successive secondary layers of cellulose are formed within the original primary wall. Once the cell structure is complete, a relatively rapid process occurring within the original growth season, the much slower process of lignification commences. This consists of the progressive deposition of lignin, initially within the middle lamella, an amorphous undifferentiated region between the cells, but then within the cellulose cell walls. This lignification serves to stiffen and strengthen the cells, occurring progressively whilst the cells remain alive. The sapwood or living xylem cells consist of a reasonably constant number of rings or depth of xylem, presumably controlled by the availability of food and oxygen to maintain the living cell processes, so that further annual growth at the cambium is accompanied at the inner surface of the sapwood by the death of cells and their conversion into heartwood. The only significant difference between sapwood and heartwood is the large amount of material that is deposited in the latter, apparently waste arising from the living processes of the tree. These deposits in the heartwood cells reduce their porosity and are often significantly toxic, so that heartwood is usually more resistant to insect and fungal attack than sapwood. These deposits also tend to make heartwood more stable, so that it is much more resistant to swelling and shrinkage with changes in moisture content.

Softwoods and hardwoods

There are fundamental differences in the nature of the wood of the conifers or softwoods and the dicotyledons or hardwoods. In softwoods the principal longitudinal cells are known as tracheids and serve both conducting and mechanical support functions. In transverse section a piece of wood appears as a honeycomb, with the annual rings arising through the change in density of appearance between the early wood with its large thin-walled cells and the late wood with its small thick-walled cell. The transverse section may also show occasional vertical resin canals whilst the radial and tangential sections may show horizontal features such as rays. In contrast the hardwoods possess fibres, similar to softwood tracheids, to provide structural support but the conducting cells are termed vessels or pores. These vessels are distributed singularly or in clusters throughout the wood, or in radially or tangentially orientated groups. For example, tangential distribution results in the ringporous woods, a term that results from the observation that a distinct ring can be seen with the naked eye on a transverse section of species of this type such as oak, ash or elm. Whilst the transverse and radial sections of both hardwoods and softwoods show annual rings which may appear to be superficially similar, they actually result from rather different variations in the structure.

In order to examine these microscopic features of a piece of wood it is necessary either to macerate the sample or to prepare thin sections. For maceration the sample is first treated with chromic acid in order to dissolve the middle lamella and thus release the individual components. Thin sections are prepared by soaking the wood until it is soft and then cutting the sections with a microtome. In both cases stains can be used which have an affinity for various individual components so that they are more readily visible under the microscope. Maceration has the advantage that it is possible to examine individual components in their entirety, but the disadvantage that their relative positions are completely unknown. Thin sections give an indication of the relative positions but it is necessary to prepare a large number of serial sections in order to encounter individual components and thus construct a three-dimensional picture of the entire wood.

This description is not intended to be a comprehensive account of the microscopic features of wood but simply a contribution to the understanding of wood structure. As an example Scots pine, *Pinus sylvestris*, has been selected to represent softwoods with European oak, *Quercus robur*, to represent hardwoods (Fig. 1.3).

The maceration of Scots pine gives a large number of thin needlelike units about 0.8 mm (1/32 in) long. These principal longitudinal structural elements are hollow, four-sided and pointed at each end. 'Pits' are scattered along opposite sides of the tracheids; the bordered pits are circular whilst the simple pits are rectangular. It is also possible to distinguish parenchyma cells in the macerated sample, each oblong, box-shaped and about 0.1 mm (1/200 in) long. Examination of sections shows that Scots pine is composed predominantly of tracheids which were laid down in regular rows as they were formed by the cambium at the outer limit of the sapwood. The tracheids are fitted end-to-end with an overlap to give both strength and continuity through the pits in the longitudinal conduction of fluids. The spring wood tracheids are large in cross-section and thin-walled compared with the summer wood tracheids which are distinguishable by their very thick walls. The only other longitudinal features are vertical resin canals which appear as spots in cross-section and as fine white hair-lines in the radial and tangential sections. These resin canals are narrow tunnels lined with small rectangular cells. The other principal features are

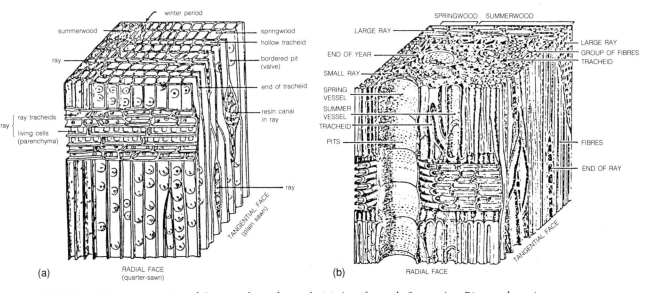

FIGURE 1.3 Reconstruction of 3 mm cubes of wood. (a) A softwood, Scots pine *Pinus sylvestris*. (b) A hardwood, European oak *Quercus robur*.

the medullary rays running as horizontal ribbons in the radial direction; with luck they might be visible in a radial section but the ends of the rays will certainly appear in a tangential section. In depth they consist of three to ten small oblong cells with their length in the horizontal direction. The top and bottom rows consist of ray tracheids with walls of irregular thickness whilst the middle rows are parenchyma cells which are connected to the vertical tracheids via the simple pits. The ray may also incorporate horizontal resin canals.

The structure of a hardwood such as European oak is entirely different. In most hardwoods the vessels are the dominant features and in cross-section they appear in the oak as large pores. These vessels run for a distance of several metres vertically within the tree and consist of the many squat tubular cells that can be seen in a macerated preparation. These cells possess thin walls so that increasing porosity necessarily leads to decreasing strength in a hardwood. The tubular cells also possess numerous pits connecting with the adjacent tracheids and fibres. The

tracheids are rather like those in softwoods but less regular in form. Fibres occur in clumps and are responsible for the principal longitudinal strength of the wood. Each fibre is spindle-shaped, long, thin and tough with a thick wall and only a small cavity. The fibres are interlocked or cemented to each other to give a hard tough wood. In addition there are small vertical rays and very large horizontal medullary rays, usually composed entirely of regular-sized parenchyma cells without the ray tracheids or resin canals associated with medullary rays in softwoods. In oak there are two sizes of medullary ray, one broad and the other narrow and barely visible to the naked eye. These medullary rays are a source of weakness as the vertical tracheids and fibres are deflected around them so that shrinkage in oak and other hardwoods is often associated with splitting through the medullary rays. In some hardwoods the medullary rays give a regular pattern, the ripple marks or silver figure frequently seen in quarter-sawn oak.

One of the features of the hardwood vessels

17

is the formation of tyloses as the sapwood is converted into heartwood. Living cells bulge through the pits to give the appearance initially of small balloons in the vessel cavities. These grow until eventually the vessel is completely blocked. Tyloses occur only in certain species; white oak possesses tyloses which block the vessels so that it is particularly suitable for barrel making, whereas red oak has no tyloses and smoke can actually be blown through the vessels.

Annual rings

The annual rings represent the amount of wood formed each year but the structure in softwood is entirely different from that in hardwoods. In the softwoods a wide ring of thin-walled spring cells is formed, followed as the season progresses by a narrower ring of thick-walled summer or late-wood cells. In hardwoods the spring wood is formed with very large vessels but as the season progresses the vessels become smaller with an increasing density of fibres and a smaller number of tracheids. In fast-grown softwoods the wood is generally of low density and inferior quality, whereas in fast-grown hardwoods the wood tends to have a high density and superior quality. This is, of course, only a generalization; very slow-grown softwoods have excellent workability but inferior strength due to their comparatively short tracheids, whilst very fast-grown hardwoods tend to lack the durability associated with slower grown wood.

Wood structure

These are the various microscopic features which are visible under a normal light microscope but it is also necessary to consider the sub-microscopic features which can only be seen with an electron microscope, as well as the ultimate chemical structure, in order to understand some of the features of wood decay and the explanations for the action of the more sophisticated wood preservative. The microscopic fibres or tracheids consist of orientated microfibrils and

it has been deduced that these in turn consist of bundles of cellulose chains. Crystalline and amorphous cellulose structures have been identified, as well as related carbohydrates such as hemicellulose and starch, all these components being assembled from sugar molecules. All the characteristic features of wood are found to be derived from these sugar-based structures and it is therefore hardly surprising that the entire purpose of the tree is to support and supply the leaves where sugars are generated through photosynthesis.

The trunk or stem of the tree is, of course, the wood of commerce. The initial twig is represented by the central pith which is surrounded by the heartwood consisting of cells which are so far removed from the bark that they have died and become filled with various extractives. Around the heartwood is the sapwood of living cells, the inner zone conducting water upwards to the leaves and the outer or phloem conveying sugars from the leaves to the living cells throughout the tree, providing them with energy to sustain life and sugar components to form cellulose, hemicellulose and starch. The outer sapwood cells immediately beneath the phloem are known as the cambium and are distinctive as they can divide to form new cells. The cambium consists essentially of two types of dividing cell, the fusiform initials and the ray initials. The fusiform initials adjacent to the sapwood give rise to xylem or wood cells which ultimately become the tracheids of conifers or softwoods and the fibres of dicotyledons or hardwoods. The outer fusiform initials give rise to the bark. The ray initials also produce xylem but in the form of parenchyma or ray cells. In this way the wood is formed so that the vertical tracheids and fibres give longitudinal strength, as well as vertical transport routes within the tracheids in conifers and within the pores surrounded by fibres in dicotyledons. In contrast, the ray cells are orientated along horizontal radial paths in order to provide conducting tissue between the phloem and the cells deep within the sapwood and heartwood, apparently conducting extractives towards the heartwood and sugars to the living cells in the sapwood, and also often providing tissue for the storage of starch, particularly in deciduous hardwoods which shed their leaves in winter and thus require a source of stored energy to enable them to survive.

The strength properties of wood can be attributed to the principal components, particularly the basic longitudinal cellular elements which are the tracheids in softwoods and the fibres in hardwoods. When a fusiform initial divides the resultant new xylem cell is surrounded by an amorphous undifferentiated substance which subsequently becomes the middle lamella. The cell rapidly achieves its ultimate length and rectangular cross-section, squeezing the middle lamella to form a thin layer between adjacent rectangular cells. The initial or primary cell wall (P layer, Fig. 1.4) consists of loose and irregularly orientated microfibrils, an important feature as this thin P layer must be capable of extension as the cell grows to its ultimate dimensions. The microfibrils are orientated in a predominantly shallow spiral, perhaps at about 60° to the vertical axis of the cell. Once the P layer has achieved its ultimate dimensions the secondary wall (S layer) is commenced and is formed typically in three separate stages. The first secondary wall (S_1 layer) is thin with a predominantly shallow

microfibril spiral, perhaps at about 50° to the longitudinal axis of the cell. The S_1 layer sometimes consists of two or more lamellae spiralling in opposite directions and is morphologically and structurally intermediate between the P and S_2 layers. The second secondary wall (S_2 layer) consists of very regular and closely packed microfibrils at a very steep spiral angle, perhaps only 10–20° relative to the longitudinal axis, and it is also it is also very thick and the dominant cell wall. Finally a third secondary wall (S_3 layer) is sometimes formed consisting of a thin shallow orientated layer of microfibrils, perhaps at about 50° to the longitudinal axis. In all cases the secondary walls are more regularly orientated than the primary wall.

These cell walls account for the cellulose which comprises about 60% of the wood substance. The S_2 layer is always dominant and it is therefore hardly surprising to find that the basic longitudinal orientation of the microfibrils, and thus the cellulose chains, within this layer account for many of the basic physical properties of wood. This S_2 layer is also rather more massive in late wood than in early wood, again accounting for the different properties between these zones of the annual rings.

The rectangular fibres and tracheids that are so readily observed when a cross-section of a piece of wood is examined under the microscope are comparatively large but their component microfibrils are quite small. It is difficult to imagine microfibrils without a knowledge of the Angstrom (Å), the unit of dimension that must be used at this scale. By definition an Angstrom is 1×10^{-8} cm, or 0.0000001 mm. Once the definition of the Angstrom is appreciated it can be said that a microfibril consists of elementary fibrils having a diameter of about 35 Å, so that microfibrils have typical diameters of 35, 70, 105, 140, etc. Å, although some microfibrils are flattened with dimensions such as 100×50 Å. If these measurements are now converted to something familiar, such as a fraction of a millimetre, it will be appreciated that the elementary

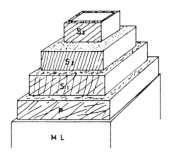

FIGURE 1.4 Cell wall layers in a softwood tracheid.

fibrils are very small, yet each consists of a bunch of about 40 cellulose chains, and thus the basic structure of the wood cell has been reduced to its ultimate chemical components.

It is usually considered that the cellulose chains consist of between 200 and 2000 glucose units, although it is sometimes reported that as many as 15 000 units may be involved. As each glucose unit has a length of about 5 Å the chains are relatively long, perhaps 10 000 Å (0.001 mm) or perhaps far longer. As wood is a natural material some discontinuity of the structure occurs but the chains lie parallel for considerable lengths, perhaps over 120 units (600 Å) or more, and it is these essentially parallel cellulose chains in the dominant S_2 layer that account for the principal properties of the wood.

Figure 1.5 illustrates the way in which glucose units are assembled to form cellulose and thus the principal structural elements of wood. This structure has considerable importance in determining the properties of wood. Firstly the sugar units are formed from water and carbon dioxide within the leaves by the process known as photosynthesis:

$$6H_2O + 6CO_2 \rightarrow C_6H_{12}O_6 + 6O_2$$
water   cabon dioxide   glucose   oxygen

The water is obtained from the surrounding soil by the roots and is conveyed up the tree through the living xylem cells in the inner sapwood to the leaves. Carbon dioxide is then absorbed from the atmosphere by the leaves, glucose is formed by photosynthesis and conveyed down the tree in the phloem between the xylem and the bark. Production of glucose by photosynthesis can occur only in the presence of the catalyst chlorophyll, the green pigment in leaves, and is dependent upon the absorption by the leaves of light energy, particularly ultraviolet light.

Energy in wood

The glucose product thus has a far higher energy level than the water and carbon dioxide constituents, and this energy can be released in a variety of ways. For example, animals eat glucose or other sugars, converting them back to the original water and carbon dioxide by the addition of oxygen and releasing energy in the process. This energy is used for maintaining life processes or for generating heat, and the burning of sugars is perhaps the most obvious illustration of the way in which oxygen can be combined with glucose to reverse photosynthesis and release energy. In the formation of cellulose chains a small proportion of the energy is lost but a considerable amount remains, and this explains why wood burns and why it is attractive to some insects and fungi as a source of nourishment.

The glucose produced by a tree is largely used for formation of structural cellulose but the enormous mass of living cells in the roots, sapwood and leaves all requires energy to maintain life, this being obtained from the glucose and associated sugars such as xylose which are also produced by the leaves. During darkness or the winter months the leaves are not producing sugars, yet the cells still require energy and obtain this from glucose accumulated

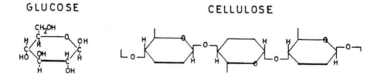

FIGURE 1.5 Glucose and the formation of cellulose.

within them or, for longer periods such as the winter, from deposits of starch which is formed from glucose and fairly readily reconverted when required. Some attacking insects are unable to utilize cellulose as a source of nourishment but they will attack wood just for the starch or simple sugar content; even mammals such as rabbits and squirrels will strip the bark from trees to gain access to the sugary sap in the phloem.

Water and wood

It has already been explained that the cellulose chains are assembled into microfibrils which are orientated in a predominantly longitudinal direction within the cell walls of softwood tracheids and hardwood fibres. The physical properties of wood largely result from this longitudinal orientation of the cellulose chains, and this is particularly the case in the relationship between wood and water. Each glucose unit in a cellulose chain possesses three hydroxyl groups which have an affinity for water. This ensures firstly that cellulose chains will wet easily but in addition water will be held between the chains, pushing them apart as illustrated in Fig. 1.6. As the chains become separated by water the bond between them becomes weaker, so that a high moisture content in wood is associated with loss of strength, particularly a loss of sheer strength and rigidity, so that a beam is more flexible and wood will cleave more readily when wet. If this separation of the cellulose chains were permitted to continue indefinitely the wood would eventually break down into individual cellulose chains and thus disintegrate, but movement with moisture change can be largely attributed to the predominantly longitudinal orientation of the microfibrils and the cellulose chains in the S_2 layer which accounts for most of the mass of the cell wall, but separation of these chains is limited by the chains in the P and S_1 layers that are wrapped around them.

As orientation of the microfibrils in the S_2 layer steepens from the pith to the bark of the tree the cross-sectional movement with change of moisture content increases approximately in proportion to the cosine of the angle of orientation. In addition, it would be expected that the longitudinal movement would be proportional to the sine of the angle of orientation, so that an angle of 10–20° to the longitudinal direction would suggest a longitudinal movement of 17–36% of the cross-sectional movement. In fact the longitudinal movement is less than half this calculated figure, apparently due to the restraining influence of the planes of lignin in the middle lamella. The middle lamella appears to have much less control on the more massive cross-sectional movement, perhaps largely because it is orientated then as a thin envelope over the swelling material in contrast with the vertical tubes in which it obstructs longitudinal movement. However, the cross-sectional movement is still restrained when it reaches the fibre saturation point, but this can be attributed largely to the very shallow microfibril angle in the P and S_1 layers which physically restrain and prevent further swelling.

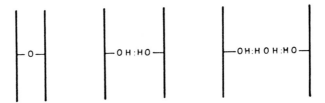

FIGURE 1.6 Water absorption forcing cellulose chains apart.

Prolonged waterlogging introduces slow but progressive hydrolysis of the cellulose and thus weakening, which eventually permits the S_2 layer to absorb water beyond the original fibre saturation point and swell to a greater extent; this is the explanation for the weakening observed in archaeological wood which has been water logged for many centuries. Careful microscopic observation discloses that water logging permits ballooning where the S_2 layer is apparently bursting through the restraining P and S_1 layers; this damage is also observed when wood is soaked with more powerful swelling solvents such as alcohols.

The loss or gain of water between the cellulose chains does not occur instantaneously with changes in the surrounding relative humidity but tends to lag. The reason is that changes will occur only under the influence of excess energy. In fact, energy is liberated when cellulose is wetted and this becomes apparent as the heat of wetting. This is virtually immeasurable and certainly insignificant in most circumstances, but it means that wetting will proceed only if this heat is removed and drying will proceed only if heat is provided. Whilst the dimensions of a piece of wood will depend on its moisture content, this lagging effect or hysteresis will mean that the wood is larger than might be expected during the drying stages and smaller than might be expected during the wetting stages, as shown in the hysteresis diagram in Fig. 1.7.

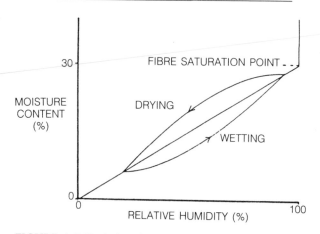

FIGURE 1.7 Variation in moisture content of wood with changes in atmospheric relative humidity.

Only 1% of the mass of wood consists of extractives, extraneous materials and minerals such as silica, the remainder being lignin and holocellulose or the carbohydrates and related compounds derived from sugar units. The crystalline structural cellulose which is the principal component of the microfibril is known as α-cellulose and accounts for abut 50% of the mass of softwoods and temperate hardwoods, the remaining holocellulose, principally hemicellulose, contributing about 25%. Lignin usually contributes about 25%, although in tropical hardwoods the lignin content is often far higher, displacing the holocellulose in proportion and giving greater rigidity but less resilience.

Wood degradation

2.1 Introduction

This book is entitled *Wood Preservation*. It is not therefore intended that it should contain a comprehensive and detailed account of wood degradation as this is not essential to the study of preservation chemistry and technology. However, the action mechanisms of various preservatives cannot be fully understood without a basic knowledge of deterioration processes, and it is obviously desirable that the most important causes of deterioration should be understood.

2.2 Biodegradation

It must be accepted that wood is perishable. Indeed, if this were not the case our forests would soon be cluttered with the useless skeletons of dead trees. Unfortunately the various wood-destroying insects and fungi are unable to distinguish between forest waste and wood in useful service. In the past wood was used principally in forest areas where it was readily available. Degrade was accepted but replacement was comparatively simple and inexpensive. In contrast, wood is now a valuable commodity and transported considerable distances between the production forest and the ultimate user. It is essential for wood to be utilized efficiently in order to conserve world resources but also to avoid unnecessary cost, both to the individual user and to importing nations as a whole.

Wood quality

Wood has been an article of commerce for many centuries, some areas exporting exotic decorative woods and others supplying straight and strong structural woods. In the past high transportation costs were justified only for the most valuable woods and even normal construction wood imported, for example, into Great Britain in the 19th century was generally slow-grown and free from sapwood. Dwindling resources have since resulted in the introduction of thinning to encourage maximum yield, giving rapidly grown trees with wide rings, principally composed of spring wood. As the trees are felled when comparatively small in diameter a large proportion of the wood is now sapwood; Swedish redwood (Scots pine) supplied as sawnwood for building is now approximately 50% sapwood. Whilst the strength properties are not significantly affected by wide rings and the presence of a high proportion of sapwood, the wood is far less durable and far less stable than the slow-growing heartwood that was commonly used in the past.

Whilst fibre and moisture content changes must be recognized as causes of degrade in wood, it is the various living organisms, both microbiological and animal, which are usually more important. The tree in the forest passes through various stages, originating with the death of the living tissue and passing through a drying stage until the protective bark is lost so that rewetting is then able to occur. Commercial wood passes through similar stages and it is possible to establish a sequence of degrading organisms to which wood is susceptible at the various stages in its progress from the living tree to the wood in service and then beyond to the various stages of destruction.

Green wood

Sapstain

Freshly felled green wood has a very high moisture content, 60–200% for most species but as high as 400% in extreme cases. Sugars and starch are present, particularly in the phloem and parenchyma tissue, and these conditions are particularly attractive to the sapstain fungi such as *Ceratostomella* species. The hyphae, the minute strands comprising the growing tissue of the fungus, are brownish in colour but when wood is affected by a mass of these hyphae light diffraction gives it a blue or black colouration, often known as blue staining or blueing. Sapstain problems can be largely avoided in coniferous woods by winter felling, which ensures a lower moisture content. Floating logs can be the cause of staining problems and, with wet wood, an increase in temperature during shipment can cause sweating if the cells are still alive and oozing of sugar-rich sap from the ends of logs followed by the development of a luxuriant fungal growth. Sapstain damage is relatively unimportant as it does not destroy the wood cell or cause any structural damage but it is unsightly and leads to downgrading and thus reduced value.

In recent years the excessive demand for softwood has caused the felling period in the coniferous forests to be extended throughout the whole year instead of being confined to the winter months alone. Wood felled during the active growing season, particularly the early spring, contains both higher moisture and sugar contents than winter-felled wood and there is a corresponding increase in the danger of staining. Indeed staining is virtually inevitable for

fellings in late spring or early summer unless they are converted and kiln-dried immediately. Even then there is still a danger that stain will develop during the early stages of kilning before the moisture content has been significantly reduced. The best solution to the problem is to convert the logs immediately into sawnwood which is then given a stain-control treatment. Alternatively the sawnwood is transferred immediately to kilns for drying and a toxic treatment is introduced into the kiln sprays. Many of the most widely used stain-control treatments are significantly volatile, particularly at the high temperatures used for kiln-drying, so that much of the treatment is lost during the process. A further treatment is therefore desirable when drying is complete in order to prevent the recurrence of staining should the wood become accidentally rewetted.

In many cases this final treatment is not used but instead the wood is packaged and wrapped in plastic sheeting in order to retain the low moisture content achieved by kiln-drying and thus prevent the development of staining. In fact, packaged wood wrapped in this way can often become very seriously affected by staining of the outer sticks in contact with the sheeting, should it be exposed to wide temperature changes which can induce evaporation of moisture from the wood during warm conditions but condensation under the plastic wrapping during a subsequent cold period. For this reason complete packages are frequently dipped in a stain-control treatment before wrapping but this achieves only limited success as it can give little protection to the inner pieces within the package. This often results in *mirror image* staining (Fig. 2.1) which develops on adjacent pieces where tight contact prevents penetration of the stain control treatment. The only proper solution is to treat pieces individually before packaging and wrapping, but most of the commonly used stain control treatments are applied in water and increase the moisture content of the wood, partly defeating the object of the

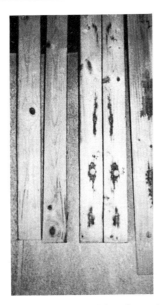

FIGURE 2.1 Mirror image staining through tight strapping during stain-control treatment. (Penarth Research International Limited)

original kiln-drying. Clearly there is scope for development of improved and more reliable stain control treatments.

Bark-borers

Bark-borers frequently attack freshly felled logs. The Wood wasps, particularly *Sirex noctilio*, attracted attention a few years ago when the Australian authorities feared that wasps introduced in softwood packaging from the northern hemisphere would damage valuable pine plantations; all wood imported into Australia is now subject to quarantine regulations to avoid the introduction of wood-borers of any type, and must be either inspected or destroyed on arrival or accompanied by a certificate to show that it has received an approved treatment. Wood wasps bore through the bark with their ovipositor and lay their eggs in contact with the phloem. The larvae hatch from the eggs and then explore the phloem, living on the sugar content of the sap and at the same time loosening the bark; this is the first stage in the destruction of a dead tree in the forest. Some Longhorn beetles behave in a very similar manner, laying their eggs in cracks in the bark. The developing larvae either explore the phloem in search of nourishment in the same way as the Wood wasps or alternatively tunnel in the sapwood, gaining nourishment from starch deposits or, in few cases, from the cellulose of the wood.

Pinhole-borers

Both Wood wasps and Longhorn beetles can be quite large, with fully grown larvae being perhaps 25 mm (1 in) or more in length. In contrast the Ambrosia beetles are much smaller, only about 3 mm (1/8 in) long. These insects, the Scolytids and Platypodids, tunnel through the bark to produce extensive galleries in which they lay their eggs. At the same time they introduce fungi which thrive on the high moisture, sugar and starch content of the freshly felled wood, thus providing the larvae with food; the browsing of the larvae on the fungus accounts for the name Ambrosia. The galleries either extend under the bark or into the sapwood, perhaps following the less dense spring wood of the annual rings, and the distinctive pattern of the galleries is frequently a characteristic of a particular species. Ambrosia attack can be very severe in the tropics, frequently introducing stain; the ambrosia fungus causes a streak of stain along the grain on either side of each gallery, even if the stain does not spread more widely. The borer damage to the wood is often described as pinholes or shotholes, depending on the size, and the damaged wood is known as pinwormy. Because of the danger of Ambrosia beetle attack coupled with staining it is normal to spray logs immediately after felling in the tropics with a mixture of a contact insecticide and a stain-control treatment, whilst sawnwood produced close to the forest before shipment also receives a further stain-control treatment as it is normally shipped with a relatively high moisture content.

Powder Post beetles

As the moisture content of the wood falls it becomes immune to fungal decay but it may be susceptible to Powder Post beetle damage. The Bostrychid adult beetles bore tunnels into the sapwood in which they lay their eggs, the hatching larvae relying on stored starch in the wood tissue for their nourishment. The Lyctids in constrast lay their eggs in large vessels or pores, although again the hatching larvae depend on the starch content in the wood. Powder Post attack is most common in hardwoods because of this starch requirement, with Lyctid attack confined to large-pored hardwoods. All wood is immune from Powder Post beetle attack if it is free from starch; temperate hardwoods are most susceptible if felled in the early winter when the starch content is at its highest level. Starch naturally degenerates slowly so that wood is virtually immune several years after felling. In addition air-seasoned wood is far less susceptible than kiln-dried, apparently because the cells remain alive during air-seasoning and thus utilize the starch, whereas the cells are killed in kiln-drying so that the starch deposits remain and attract Powder Post beetle attack.

Furniture beetles

The Anobids or Furniture beetles, are a very important family. The bark-borer *Ernobius mollis* lays eggs in cracks in the bark of dry softwood logs or bark still attached to the waney edges of sawnwood. The hatching larvae explore the phloem in search of nourishment and loosening the bark, but the galleries sometimes extend for a distance of up to 12 mm (1/2 in) into the sapwood. These holes in the sapwood are sometimes confused with those of the Common Furniture beetle *Anobium punctatum* but they tend to be rather larger in diameter and they are invariably associated with galleries under the bark; the bark may have fallen away from the adjacent waney edge but the galleries can still be recognized. Confusion with the

Common Furniture beetle is the most important aspect of *Ernobius mollis* attack. It is dependent on sugar in the phloem and to a lesser extent starch in the sapwood. This deteriorates naturally a year or two after conversion, so that *Ernobius* is eliminated naturally without any need for treatment.

If small holes are observed in the sapwood of a piece of softwood with no evidence of either bark or waney edge, then the Common Furniture beetle *Anobium punctatum* is most likely to be responsible. Common Furniture beetle attack will occur in the sapwood of most softwoods and hardwoods, as well as in the heartwood of some temperate hardwood species such as birch and beech. Eggs are laid, usually in the summer months, in cracks on the surface of the wood or on furniture in open joints. The larvae hatch by breaking through the base of the egg and boring straight into the adjacent wood. The insect usually remains in the larval stage for almost a year before forming a chamber just below the surface, in which the full grown larva about 5 mm (1/5 in) long pupates into an adult beetle. The adult bores to the surface and escapes by a circular flight hole about 1.5 mm (1/16 in) in diameter. The adults mate and the female then initiates a fresh attack on the wood by laying eggs, sometimes in the old flight holes. Obviously the damage is caused entirely by the larvae boring through the wood, in contrast to Ambrosia beetle attack in which the adults are responsible for tunnelling.

Whilst the life cyle of the Common Furniture beetle is often as short as 1 year, it can be extended if conditions are unfavourable to 4 years or more before the larvae are able to store sufficient energy to survive through the pupation stage when they develop into adult beetles. If wood becomes damp it will considerably encourage the activity of the Common Furniture beetle, enabling it to attack normally resistant heartwood. It appears that, although deterioration may not be obvious, fungal or bacterial activity is converting the wood to a form that is more readily assimilated by the insect. The Death Watch beetle *Xestobium rufovillosum* is an Anobid related to the Common Furniture beetle which is entirely dependent on the presence of microbiological attack. Indeed, wood attacked by Death Watch beetle is often brownish in colour, indicating prolonged incipient fungal attack although there may be no other evidence. The Death Watch beetle is far larger than the Common Furniture beetle and in the British lsles it is generally associated with old buildings with oak chestnut timbers but particularly historic churches, apparently because the periodic heating combines with the traditional lead roof covering to encourage condensation, incipient fungal attack and thus Death Watch beetle attack in the roof sarking boards. However, the rather sinister name of the Death Watch beetle is not associated with churches but with the characteristic tapping which is used by the adult beetles as a mating call and which is produced by striking the top of the head against the surface of the wood on which the beetle is standing. On a calm day the sound is distinctly audible in a quiet building, perhaps particularly in one that is silent through the presence of death.

House Longhorn beetle

Whilst it is logical to treat the Anobid beetles as a group progressing from the *Ernobius* bark borer to the Common Furniture beetle attacking dry wood and ultimately the Death Watch beetle dependent on some decay, this unfortunately displaces the House Longhorn beetle from its position in association with the Common Furniture beetle in this sequence of attack. The Longhorn or Cerambycid beetles are a large group of considerable importance in the forest but the House Longhorn beetle *Hylotrupes bajulus*, known in some countries as the House-borer, has adapted to dry conditions in buildings. The attack is confined to the sapwood of dry softwoods and is thus of particular importance where wood with a large sapwood content

is used for structural purposes. The life cycle is basically similar to that of the Common Furniture beetle but varies typically from 3 to 11 years. This is a large insect with a fully grown larva perhaps 30 mm (1 1/4 in) long. A female beetle may lay up to 200 eggs and, if most of these hatch within the same roof or floor structure, very substantial damage can be caused during the years before the adult beetles ultimately emerge. Indeed, the entire sapwood may be riddled by oval galleries, leaving a thin intact veneer over the surface of the piece of wood. The adult beetle emerges from the wood through an oval exit hole about 10 mm (3/8 in) across, but the presence of this flight hole is a certain indication that severe damage has already been caused and the structural integrity of the piece of wood must be checked by probing.

The Death Watch beetle is only one of several wood-borers which are dependent on damp or wet conditions. The presence of wood-boring weevils is often indicated by small holes in damp wood attacked by Death Watch beetle or Common Furniture beetle. If large pieces of relatively durable wood such as European oak remain in a damp and slowly decaying condition for a prolonged period they may attract the attention of other insects such as *Helops coeruleus*, a large blue beetle with a long larva possessing two recurved spines on its tail, or even Stag beetles which are more commonly found in decayed roots or fence posts below ground level. These insects are, of course, dependent on fungal decay which is a far more significant factor in wood degradation than the insect damage.

Fungal decay

Fungal damage can originate within the standing tree. Whilst the very high moisture content in the living sapwood tissue generally prevents fungal decay and insect attack, physical damage to the bark and cambium or scars from the removal of branches may permit the moisture content to fall to a level at which fungal decay is possible. Branch and root scars are particular danger points as they may expose relatively dry heartwood which, whilst it is relatively durable in most species, will permit decay in others such as beech. In the case of a branch scar the first stage is probably rapid drying from the exposed end-grain and the development of checks or splits. Subsequent rainfall is then trapped within these cracks and produces a gradient of moisture content which permits spores in the atmosphere to encounter precisely the conditions that they require for germination. The fungal infection then spreads progressively through the heart of the tree.

Dote

These heart rots are variously known as dote or punk, or they are referred to by the name of the attacking fungus, such as Honey fungus *Armillaria mellea*. There are no obvious external signs of damage until the decay is far advanced and a sporophore or fruiting body appears through the bark, usually indicating that the heartwood of the tree is virtually hollow. In doty heart the attack is widespread but damage is limited perhaps because the heartwood moisture content is too low. Whilst the infection will be killed if the wood is kiln-dried at a sufficiently high temperature, failure to take this precaution introduces the danger that the infection will become reactivated if the moisture content reaches an adequate level. For example, logs are frequently stacked in the open and if they are doty they are likely to become covered with the fluffy white hyphae or growing elements of the reactiviated fungus. Washing floors can also reactivate dote, the upper surfaces softening progressively until they eventually break up and fail. Dote normally causes a colour change in the wood, most frequently to a brown colouration in the centre of the heart, although occasionally off-centre, when the effect is referred to as false heart. There is usually a smell characteristic of the particular fungus involved.

Brown and White rots

The only special feature about dote is its origin in the standing tree or in a log that has been permitted to remain for a protracted period in the forest. In all other respects dote is characteristic of the most important group of wood-destroying fungi, the Basidiomycetes. In the case of Brown rots the fungus destroys the cellulose, leaving the lignin which gives the wood a characteristic brown colouration and usually cross-grain cracking. Common dote, *Trametes serialis*, is a Brown rot but other dotes are described as White rots because the fungus decays both cellulose and lignin causing less colour change and a fibrous appearance. *Fomes annosus* is a dote found in spruce, larch and red oak heartwood, giving a purplish colouration. This colour is particularly intense in spruce and lighter in larch. The attack is typical of dote, appearing on the cut surface as tiny white pockets of rot filled with growth like small pieces of cotton-wool. If dote is suspected but not apparent in this way it can usually be detected by lifting the fibres with a knife; if they are long and springy the wood is sound but decay must be suspected if they are brash and break readily across the grain.

Any dote activity that redevelops in wood in service can be ultimately attributed to the germination of a spore, perhaps on a branch scar on a standing tree or a log lying in the forest. The other principal decays of wood in service differ only in the fact that spore germination occurs on the surface of the wood in service, when the conditions result in an appropriate moisture content within the wood and relative humidity in the surrounding atmosphere. A vast selection of spores is invariably present in the atmosphere and a fungal infection will inevitably develop which is appropriate to the conditions and the wood species. For example, a post standing in moist ground will generally provide all conditions varying between the high moisture content in the ground and the low moisture content in the well-ventilated aerial parts but it is the intermediate conditions at the ground line that prompt the most serious decay; in the forest this ground line damage ensures that the relatively dry tree falls to the ground where undergrowth inhibits evaporation and the increasing moisture content then causes decay in previously dry parts of the tree.

Wet rots

There are a very large number of wood-destroying fungi that occur throughout the world. However, it must be emphasized at this point that there is never any need for decay in wood used in building as simple design precautions are usually adequate to ensure that the wood never becomes sufficiently damp to support decay fungi. There are some special conditions, such as external window frames, piles in the ground and frame walls and roofs in which condensation can cause wetting, where decay is highly probable but it can be avoided by taking proper precautions such as the selection of suitable durable wood or the use of an adequate preservative treatment.

Fungal decay frequently occurs in buildings despite the ease with which it can be avoided, usually through neglect in design or maintenance. The Cellar rot, *Coniphora puteana*, is a particularly good example of a decay that can occur as a result of neglect. It occurs in persistently damp conditions, such as when a damp-proof course is omitted and when plates under floor joists are in direct contact with damp supporting walls. If the moisture content tends to fluctuate as in wood affected by a periodic roof leak, the White Pore fungus, *Poria vaporaria*, is far more common in softwoods whilst other fungi occur in hardwoods, such as the Stringy Oak rot, *Phellinus megaloporus*, in oak. *Coriolus versicolor* sometimes develops when

non-durable tropical hardwoods are used for drips or sills on external joinery (millwork), and *Paxillus panuoides* generally occurs where the conditions are too wet for the Cellar fungus.

These are only a few of the fungi that may be encountered but they serve to illustrate the way in which a fungus is often associated with a particular combination of wood species and conditions. The *Coniophora*, *Poria* and *Paxillus* species are all Brown rots giving affected wood a distinct brown colouration and a varying degree of cross-grain cracking. The *Phellinus* and *Polystictus* species are White rots causing only limited change in the colour of the wood but very pronounced softening and loss of strength.

Dry rot

Perhaps the fungus that is most widely known as causing serious damage in buildings is the Dry rot fungus *Serpula lacrymans*. Dry rot spores will germinate only when the atmospheric relative humidity is suitable. Usually accidental wetting, perhaps from a plumbing leak or a roof defect, has caused wood to become very wet and subsequent slow drying, perhaps a seasonal effect or through correction of the defect, permits the wood moisture content and the relative humidity in contact with the surface to reduce slowly through the optimum condition for spore germination. Spore germination consists of the development of hyphae or threads which penetrate into the wood, radiating from the original point of germination and branching so that the affected area is covered with a soft white growth like cottonwool. If favourable and unfavourable conditions alternate successive masses of hyphae will become compacted on the surface of the wood to form dense skin or mycelium.

The growth will not be confined to wood but will spread across and through plaster, brickwork, masonry and concrete in an attempt to discover further supplies of wood for nourishment. This exploratory growth must be provided with food from adjacent wood attacked by the fungus and the hyphae develop into rhizomorphs or conducting strands which convey food as well as water absorbed from damp masonry. The fungus will use this water and water formed during the destruction of cellulose to form globules on the surface of the growth which, if ventilation is restricted, will maintain the atmospheric relative humidity at the optimum level for growth. This habit of forming 'tears' on the surface of the growth explains the Latin name *lacyrmans* and the description in the Old Testament Book of Leviticus, Chapter XIV, of Dry rot as the 'fretting leprosy of the house'. As an attack progresses the cellulose is destroyed, giving the wood the characteristic dark brown colour of a brown rot, accompanied in the case of Dry rot by very pronounced cross-grain cracking which, in combination with longitudinal cracking, gives the decayed wood a characteristic cuboidal appearance. The preference of Dry rot for unventilated situations in which it can control the humidity ensures that it generally remains largely concealed and the first sign of damage other than a characteristic odour is perhaps the buckling and cracking of a painted skirting. By then the attack may well be very extensive, perhaps spreading through masonry for considerable distances in all directions in the search for wood. When the wood supply is nearing exhaustion or when the humidity falls unexpectedly the fungus may spread onto the surface of the concealing wood or plaster and form a sporophore or fruiting body which will produce millions of red-brown spores in an attempt to infect any other wood that may be in suitable condition in the vicinity.

Soft rot

Wood that is continuously immersed in water becomes saturated and immune to the attack of Basidiomycetes, the Brown and White rots, but Soft rot can occur which is caused by Ascomycetes and the Imperfect fungi. Soft rot takes the form of a softened layer of wood on all exposed surfaces, the damage progressively increasing in depth at a very slow rate. Unprotected wood immersed in fresh or sea water is invariably affected in this way but the damage is comparatively insignificant in large sections such as those used for piling and the construction of groynes. In wood of thinner section the loss in strength can be significant, as in a neglected boat where planking has been exposed through abrasion damage to the paintwork, or in cooling-tower slats where the high temperature results in particularly rapid Soft rot attack, even in the presence of some preservative treatments.

Marine-borers

Continuous immersion in sea water also introduces the danger of marine-borer attack. Several animals can infest wood in this way but damage is most commonly due to crustaceans called Gribble (*Limnoria* species) and molluscs called Shipworm (*Teredo* species). Gribble attack takes the form of superficial tunnelling into the wood. Small holes less than 2.5 mm (1/10 in) in diameter are produced and intensive attacks seriously weaken the surface of the wood which progressively erodes, exposing fresh wood to attack. Gribble is a major marine problem in most parts of the world. In contrast, shipworm attack is very difficult to detect. The animal enters the wood by boring a hole about 0.5 mm (1/50 in) in diameter. This is then extended to form a long and progressively widening tunnel with a characteristic calcareous lining. Severe attack may considerably reduce the strength of the wood but this may not become apparent unless abrasion or Gribble attack exposes the tunnels and their characteristic white lining. Shipworm is generally confined to relatively warm and saline water so that in Europe it is not found in the north except on coasts warmed by the Gulf Stream and it is absent from river estuaries and the Baltic Sea where the water is insufficiently saline.

In boats most normal anti-fouling paints provide efficient protection against marine-borer attack but in areas where marine-borers are a particular problem regular inspections are advisable to ensure that unprotected wood is not exposed by abrasion. For example, wooden rudder trunkings cannot be effectively coated with anti-fouling paint and shipworm frequently becomes established as a result. Paint cannot be applied regularly to heavy marine woodwork such as piles, fenders and groynes, so that naturally durable or adequately preserved wood must be used. Wood is relatively easy to preserve against shipworm attack but gribble is a far greater problem; most preservatives will reduce the rate of attack but will not ensure complete protection.

If wood is waterlogged and completely surrounded by an impervious clay or mud layer there is no possibility of fungal decay or borer attack because of the complete lack of oxygen. However, some anaerobic bacteria are able to survive in these conditions, obtaining the necessary oxygen by their reducing action on suitable chemicals that may be present, such as by the conversion of sulphate to sulphide and thus the generation of the hydrogen sulphide odour which is a typical feature of these conditions. These bacteria do not cause any obvious damage to wood, although it becomes very dark brown in colour, but protracted waterlogging results in slow hydrolysis of the cellulose, which is thus progressively lost. The amount of loss can be assessed by determining the dry density of the wood but this is very difficult as any attempt to dry the sample invariably results in considerable distortion, often rather reminiscent of the shrinkage that occurs from Brown rot attack which is, of course, another method for removing the cellulose from wood and leaving the brown lignin. The alternative is to measure the saturated moisture content of the sample by weighing firstly when wet and subsequently after over-drying. The amount of water loss is then related to the dry mass to give the saturated

moisture content and this can be directly related to the period of immersion. This technique is sometimes used by archaeologists to estimate the age of an ancient structure when they are able to obtain waterlogged wood samples from, for example, a sunken boat or piles driven into a bed clay.

Termites

Termites or White Ants are probably the most serious wood-destroying pests. They are not ants but belong to the Isoptera whereas true ants are Hymenoptera, an order which also includes the bees and the wasps. The termites are social insects like the true ants, living in communities with specialized forms or castes, the workers and soldiers, as well as male and female reproductive individuals. Termites are widespread in tropical and sub-tropical countries, and the rate and severity of their attack make them a serious pest and economically significant wherever they occur. There are approximately 1900 identified species of termites and more than 150 are known to damage wood in buildings and other structures. With such a vast number of species involved it is essential when considering preservation systems to have a knowledge of their basic behaviour and differences between the various families.

Termites are principally tropical in distribution but are encountered as far south as Australia and New Zealand and as far north as France and Canada. Improvements in transportation and world trade have been responsible for the wider distribution of termites and this is particularly apparent in France and Germany. The termite of Saintonge, *Reticulitermes santonensis*, is native to the west coast of France between the rivers Garonne and Loire, but it has spread to Paris where it is concentrated around the Austerlitz station which serves this coastal region. The North American species *Reticulitermes flavipes* occurs in Hamburg where it was introduced in ship cargoes. Both these species are sensitive to climatic conditions and are not expected to spread widely; they are largely concentrated in heating pipe ducts and other rather warm areas in buildings. However, it is still remarkable that the British Isles remain free from termites.

A common feature of the six families of termites within the order Isoptera is the lack of cellulase in their digestive enzymes, despite the fact that all six families possess members which are wood-destroying. Three of the families are of only limited significance; the Mastotermitidae are represented by only a single primitive species in northern Australia, the Termopsidae include three species that infest buildings in North America, and the Hodotermitidae are confined to semi-desert areas of South Africa, North Africa and the Middle East. The remaining three families are of considerable significance as they contain the more important wood-destroying termites.

The family Termitidae is a large and mixed group which includes the subterranean and the mound termites, which construct nests under the ground, on the sides of trees or as mounds on the ground. They are able to digest wood without the assistance of an intestinal symbiont because they rely on fungus for the production of a cellulase to convert the wood to a form suitable for their own digestion. The Termitidae can be divided into two distinct groups. The first group contains the *Microcerotermes*, *Amitermes* and *Nasutitermes* species which depend on prior infection by a fungus to convert cellulose to a digestive form. Lignin is not digested and is passed through the gut, providing the raw material for the construction of the typical honeycomb nests and covered walkways. In contrast the group containing the *Macrotermes*, *Odontotermes* and *Microtermes* species is not restricted to wood already infected by fungus but instead these termites convert all cellulose to a digestible form in fungal gardens within their nests. Fresh wood is gnawed into fragments which are then chewed to paste and placed in the nest where they become infected

by the termite with a fungus which causes deterioration of both cellulose and lignin. The wood is therefore converted to fungal tissue which is then eaten by the termites.

The Kalotermitidae family includes the Dry Wood termites which, as this name implies, are able to digest wood possessing a very low moisture content. Symbiont protozoa in the hindgut provide cellulases sufficient to enable these termites to digest wood cellulose in a normal manner, although lignin is not digested and is passed through the gut. Colonies of these termites inhabit sound dry wood and rarely enter the ground; for this reason all the other families are sometimes collectively known as subterranean termites. Dry Wood termite attack is spread by winged egg-laying females and, in areas where there is a risk of Dry Wood termite attack, it is essential to use naturally durable or adequately preserved wood.

The termites in the Rhinotermitidae family are sometimes described as the Moist Wood termites; in contrast those in the minor family Termopsidae are sometimes described as Damp Wood termites. The Rhinotermitidae cause damage to buildings and other structures but have far less economic significance than either the Termitidae or the Kalotermitidae. They possess protozoan intestinal symbionts but they prefer moist wood which is already infected by fungi or bacteria, thus achieving more effective digestion and assimilation than the Dry Wood termites.

In many areas the main termite hazard is confined to the subterranean species of the Termitidae. The first group of this family depends on prior infection of the wood by fungus and they are therefore a particularly serious hazard to fence posts, transmission poles and all other wood in soil contact. Attack can be readily prevented by taking the conventional precautions to avoid fungal decay, such as ensuring that wood remains dry in buildings and using only wood that is naturally durable or adequately preserved in ground contact. In the case of the second group control is more difficult as they will physically destroy undecayed dry wood, ingesting fragments and transporting them back to the nest where they process them in fungal gardens and thus convert them into useful food.

Physical barrier systems provide the most common method for protecting building structures from this particular hazard. The Termitidae are unable to fly and protection can be obtained by isolating wood from the soil by the use of shields of metal or plastic between the wood and the footings, by introducing a barrier of poisoned soil or concrete, or by painting structural components white if they provide a route to the wood; the termites dislike constructing their walkways on light-coloured surfaces. These barrier systems are reasonably effective provided they are conscientiously constructed to prevent the termites discovering or constructing an alternative route to the wood structure. Subterranean termites are capable of constructing unsupported tubular walkways spanning distances of 300 mm (12 in) or more and this may enable them to bridge shields. In addition, barriers of this type are completely useless if bridged by negligence in subsequent construction or maintenance, such as the careless installation of electric cables and plumbing.

Many other termites, particularly the Dry Wood termites or Kalotermitidae, are able to fly and cannot be realistically controlled by physical barrier systems. If wood is to be used in structures exposed to attack it must be naturally resistant or adequately treated with preservative if it is to survive.

2.3 Moisture content fluctuations

The properties of wood are profoundly influenced by the presence of water. The moisture content is very high in standing trees or freshly felled wood, usually known as green wood, varying typically from 60 to 200%. The green moisture content tends to vary inversely with

the normal dry density of a wood so that black ironwood with a specific gravity of 1.08 has a green moisture content of only about 26% whereas South American balsa with a specific gravity of 0.2 has a green moisture content of about 400%. During drying free moisture is first lost from the cell spaces and this involves little change in properties except for a change in density. Eventually drying will result in the wood reaching the fibre saturation point, normally at a moisture content of 25–30%, and further drying results in the loss of bound moisture from the cell walls.

Fibre saturation point

The loss of bound water reduces the separation between adjacent cellulose chains and causes shrinkage as well as progressive changes in physical properties. The amount of bound water remaining in the wood is approximately proportional to the relative humidity of the atmosphere, although changes in moisture content lag behind changes in relative humidity, a phenomenon known as hysteresis. Most of the changes in properties with variation in moisture content can be attributed to the submicroscopic or chemical structure of wood, as explained previously in Section 1.3.

Movement

The swelling or shrinkage with changes in moisture content is known as movement. In the longitudinal direction the movement is generally very low, only about 0.1% for a change of moisture content from normal dry wood at about 12% to the wet condition at fibre saturation point or above. In contrast the movement between the radial and tangential direction can be largely attributed to the fact that early or spring wood shrinks less, so that in the radial direction the movement can be attributed to the average of the early and late wood shrinkage, whereas in the tangential direction the dense and strong ribs of late wood tend to control the physical behaviour and thus generate higher movement. Indeed, if a piece of softwood is planed smooth at a low moisture content and then wetted, the late wood will swell to a greater extent than the early wood and a corrugated surface will be produced.

Generally, a higher lignin content in a particular wood species will result in lower movement or greater stability. The higher shrinkage in the tangential direction, coupled in some species with weakness induced by large medullary rays, sometimes results in surface splits developing during drying. However, it has been explained in Chapter 1 that the movement characteristics can be largely attributed to the microfibril angle in the cells, and as this changes progressively from the pith outwards there is also a progressive change in movement properties. Heartwood close to the pith is most stable but, in addition to this progressive change, there is a far larger alteration in movement between heartwood and sapwood in many species, the sapwood often possessing extremely high movement (Fig. 2.2). These changes in movement relative to the original position in the tree combine with distorted annual rings and twisted grain to cause warping and splitting, defects that will be described in more detail later in this section.

Water vapour is lost or gained most rapidly through end-grain surfaces because of their permeability, a factor which also encourages the very rapid absorption of liquid water. This means that many defects arising through moisture content changes, such as splits and the development of decay or stain fungi, are concentrated around end-grain surfaces. In general, wood is used as long pieces with only small end-grain surfaces so that the side-grain is usually more significant, particularly in service in dry conditions where it is the hygroscopicity of the wood coupled with changes in atmospheric relative humidity that are likely to have the greatest effect. Wood softens as moisture content increases towards the fibre saturation point, a property that is very valuable in some circumstances such as when bending wood or peeling veneers from logs, although it significantly reduces strength so that in structural uses wood must be employed in adequate dimensions to tolerate the strength loss that will occur should the moisture content approach the fibre saturation point.

Shrinkage in buildings

When trees are first converted to sawnwood the moisture content is significant because of the danger of fungal damage and also because of

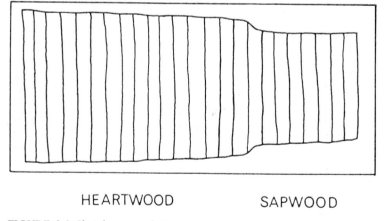

HEARTWOOD   SAPWOOD

FIGURE 2.2 Shrinkage on drying.

the excessive transport costs that are involved if the weight of wood is perhaps doubled by a high moisture content, but if this wet wood is used in buildings the most serious problem, other than the danger of fungal decay, is the cross-grain shrinkage that will occur as the wood gradually dries and achieves an equilibrium moisture content with its surroundings. This is a very serious problem in certain uses such as window and door frames, floorboarding, panelling and other situations in which cross-grain shrinkage will result in the development of unacceptable gaps between the individual pieces of wood. It is therefore usual to air-season or kiln-dry wood before installation, to the average moisture content that it will encounter in service. The main problem is that wood may be processed to the required moisture content but this may then change during storage, delivery or installation, particularly in buildings where the wet trades such as bricklaying and plastering result in very high humidities during construction. Indeed, floors are sometimes laid in new buildings very promptly after kiln-drying and the high humidity then causes them to expand so that sometimes the entire floor lifts in a dome or the expansion causes serious damage to the surrounding walls (Fig. 2.3).

FIGURE 2.3 Parquet flooring distorted through moisture absorption after being laid. (CSIRO, Australia)

Stacking the wood for a period inside the building before laying will avoid this particular problem as the wood will thus reach equilibrium with the high humidity in the damp new building but when the building is ultimately occupied the relative humidity will fall and shrinkage defects will develop. Quarter-sawn (radial cut) floorboards will simply shrink in width leaving gaps at the joints but flat-sawn (tangential cut) boards will cup in addition; they should be laid with the heart surface upwards so that the cupping causes each board to be raised in its centre rather than at its edges where it can cause tripping. The integrity of the floor and thus its ability to act as a fire-break or thermal insulator is naturally affected by this shrinkage, unless the boards are joined by tongues to span the shrinkage gaps.

Panel products

A floor is a rather large panel normally assembled out of separate boards and the problems encountered in floors are obviously encountered in other panels such as wall linings, doors and furniture. Several panel materials are now available in which wood has been processed to minimize these shrinkage defects. In plywood the wood has been cut into veneers which are orientated at right angles to one another so that the stable longitudinal grain is used to physically restrain the movement in the cross-grain direction. This method for reducing shrinkage is particularly successful in normal buildings but sometimes fails under extreme conditions such as when used in boat building. Adhesives have been developed which will withstand the very high stress that is developed when the moisture content fluctuates from fibre saturation point down to low levels in very dry conditions but if this occurs regularly there is a tendency for the wood to rupture on either side of the glue line and this can be avoided only by using wood of moderate or low movement in the manufacture of the plywood. It should be appreciated that the plywood still expands normally in

thickness but this is seldom a problem unless it affects the weather seal when plywood is used as an exterior panelling.

One particular advantage of plywood is that, as the grain runs longitudinally in both directions within the plane of the board, it is equally strong in both directions, although not as strong as the longitudinal direction for normal wood. In contrast it could be said that the alternative process for producing a panel by the manufacture of particle-board results in a material that is equally weak in both directions. In particle-board manufacture the principle is to divide the wood into small particles which are then randomly orientated so that the movement is shared equally in all directions, although it is restrained to some extent if sufficient adhesive is used. The movement is higher than for plywood, and, whereas plywood presents a completely smooth surface, a particle board is comparatively irregular, a defect that is particularly noticeable with changes in relative humidity as the surface chips expand and contract across the grain. Smaller particles are sometimes used for the surface of the board in order to minimize this problem or the surface is veneered, although veneers applied on boards with comparatively large surface chips will not conceal this movement which will show through the veneer as shadowing.

The only alternative is to reduce the wood to individual fibres which are then randomly oriented and reconstructed into a board which is, in effect, a rather thick sheet of paper. Low-density boards of this type are known as insulation boards and are used for lining walls and ceilings, but the medium- and high-density boards or hardboards present a reasonably smooth surface for painting. Whilst fibre-boards might appear to be similar to plywood in principle, although the wood components are reorientated on a rather different scale, they do not achieve the same end results. In fibre-boards the movement tends to be randomized in all directions, whereas in plywood the movement in the plane of the board is physically restrained. Plywood is therefore preferred for external panelling and other situations where it may achieve a high moisture content as movement within the panel dimension remains small, whereas a fibre-board will swell significantly in similar conditions and this may lead to serious distortion.

Seasoning

Shipping specifications often require wood to be properly seasoned for shipment to the country of destination. This generally means that the moisture content of the wood must be sufficiently low to ensure that no deterioration occurs within the hold of the ship. In theory the wood must be dried in some way to a moisture content below about 22% compared with an average moisture content of 60–200% when freshly felled. Seasoning is the term that is generally applied to the processes that are adopted for reducing the moisture content. Various shrinkage defects can develop if wood is dried too rapidly so that much of the expertise in seasoning is devoted to achieving the maximum drying rate whilst still avoiding the development of defects. Shrinkage and swelling will also occur as the moisture content varies in service so that seasoning should attempt to dry the wood to the average moisture content that it will achieve in final use, in order to minimize movement defects.

The traditional method for seasoning or drying wood is to stack it in the open air, although it will be appreciated that drying will be achieved only if the stacks are protected from rain, either by providing a roof to each individual stack or by the use of large open-sided seasoning sheds. One serious problem with air-seasoning is the excessive drying rate during hot weather. This can be a very serious problem as drying occurs most rapidly from the porous end-grain which thus shrinks in advance of the wood further along a piece, causing the development of severe

splits or seasoning checks. This defect can be partly avoided by the use of end cleats, generally pieces of wood or metal nailed across the end-grain in order to physically restrain the splitting. This method for controlling checks is very unreliable and a more efficient technique is to seal the end-grain with a bitumastic or wax formulation so that drying is confined to the side-grain throughout the length of each piece of wood. When the wood has a high moisture content in its initial green state it is comparatively flexible and it must be carefully stacked in order to avoid sagging. Stickers or piling sticks must be placed between the pieces to permit a proper circulation of air, or alternatively the pieces may be stacked with each alternate layer in a different direction, a method known as self-piling which is frequently used for the drying of round wood such as transmission poles.

In kiln-seasoning the wood is placed in containers or kilns in which the temperature and humidity can be controlled in order to achieve the maximum rate of drying consistent with freedom from the development of defects. The kiln must be designed and the wood carefully stacked so that a uniform air circulation can be achieved. If the air is always passing in the same direction it is obvious that the temperature will be lower and the relative humidity higher at the exit, so that the wood nearest the exit will have a higher moisture content than that close to the inlet. In many kilns the air flow can be reversed in order to reduce these differences. One advantage of kiln-seasoning is the high temperatures that are used towards the end of the drying sizes of the pieces of wood that are being dried when controlling a kiln, although cross-section dimensions will be important if it is desired to estimate the time to complete the drying process.

Fibre saturation point usually occurs at a moisture content of about 25–30% and shrinkage occurs in wood dried below this point. Air-drying will normally reduce the moisture content to 17–23%, and if lower moisture contents are required kiln-drying is essential. Wood can be pressure-treated with preservatives at moisture contents below 25%. Carcassing or framing timber in buildings can tolerate a moisture content of up to 23%, mainly because the cross-section dimensions in which movement occurs are relatively unimportant and drying is necessary only to achieve resistance to fungal decay. Wood intended for use in ships, boats and vehicles should be dried to 15%. In a house with reasonable central heating the average moisture content should be 12%, although bedroom furniture would perhaps be better manufactured with a moisture content of 14%, more closely related to the lower heating level in these rooms. In buildings which are more intensively heated such as offices and hospitals, a moisture content of 8% may be necessary, and perhaps less in flooring installed on top of floor heating.

In climates with cold dry winters the atmospheric relative humidity is very low and results in moisture contents as low as 4% in buildings with central heating, yet in the spring the high humidity may increase the moisture content to as much as 12% so that the choice of a preferred moisture content for kilning is rather difficult. Indeed these required moisture contents are all largely theoretical as it is very difficult to ensure that wood remains completely protected during the numerous handling and transportation stages after leaving the kiln. The worst stage is perhaps installation in a new damp building in which any wood will automatically recondition to a higher moisture content in equilibrium with the new conditions.

It has already been explained that wood shrinks when its moisture content is reduced, that the outer zones of the trunk shrink more than the inner, and that the sapwood shrinks more than the heartwood. In addition, the outer layers of a log or piece of wood are ventilated and thus dry more rapidly than the inner cycle as these eradicate any insect infestation as well as many fungal infections.

Atmospheric relative humidity

The ability of air to absorb moisture varies with the temperature so that the moisture content at which air becomes saturated increases with the temperature. When air is completely saturated it is said to have a relative humidity of 100%, and a relative humidity of 50% naturally means that the air is half saturated. If air increases in temperature the relative humidity will decrease or the air will appear to become drier as it is capable of holding a larger amount of moisture at the higher temperature, yet the actual humidity or moisture content of the air will remain unchanged. Thus the drying power of air will be increased as the relative humidity is reduced, either by reducing the actual humidity or moisture content or simply by increasing the temperature to improve the moisture holding capacity. Wood cannot dry if the relative humidity of the air is 100%. In addition, if wood is allowed to remain in air possessing a constant relative humidity, it will eventually acquire a moisture content in equilibrium with the surrounding air. The moisture content in equilibrium with 100% relative humidity is known as the fibre saturation point and lower moisture contents are in equilibrium with lower atmospheric relative humidities. It will therefore be appreciated that the actual operating temperature of a kiln is comparatively unimportant as it is the relative humidity that is responsible for the rate of drying.

Various kiln schedules have been devised to dry wood as rapidly as possible without the development of seasoning defects. Obviously the rate of drying will depend on the cross-section of the pieces of wood and it is therefore important that a kiln should always be loaded with wood of similar dimensions. In addition, kiln schedules are normally devised so that they are controlled by the moisture content of the wood at the air inlet side of the kiln. In a typical schedule it will be specified that, when this moisture content reaches a certain level, the relative humidity of the incoming air must be reduced by raising the temperature in order to continue the drying process. In this way time is ignored in kiln schedules and it is also unnecessary to consider the actual layers, a combination of all these effects leading to the distortion and tissue rupture defects which are a characteristic of faulty seasoning or kiln-drying. One particularly important point is that wood is plastic when wet but set when dry so that it is essential that wood should be carefully stacked flat to avoid unnecessary distortion.

Warping

Warping is the general term for seasoning distortion and takes several forms. Cupping is warping across the grain or width of a board and arises in flat-sawn boards through the outer zones of the trunk and particularly the sapwood shrinking to a greater extent than the inner zones, giving a planoconcave surface on the outer side of the board relative to the trunk. Flat-sawn boards used for flooring should always be laid with the heart upwards to give a planoconvex surface with the prolonged drying that will occur in service and thus avoid trip edges. The alternative is to use more expensive quarter-sawn boards which are resistant to cupping and also more hard wearing. Twisting arises in boards through the presence of spiral or interlocking grain and is really the result of wood selection rather than a seasoning defect, although spiral grain is quite common and perhaps unrealistic to reject. Bowing is longitudinal curvature arising perpendicular to the surface of a board, either through spring in a flat-sawn board or more usually through sagging in the stack during drying. Spring or crooking is longitudinal curvature within the plane of the board and can be severe in some species such as elm or kempas grown in swampy areas, giving decreased strength. Finally diamonding occurs in pieces of wood of rectangular section when the annual rings pass diagonally across the end-grain, as shown in Fig. 2.4; the

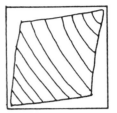

FIGURE 2.4 Diamonding or unsymetrical shrinkage in tangential and radial directions.

radial shrinkage is less than the tangential shrinkage so that the diagonals in a square section piece of wood become different in length and give a diamond shape.

Splitting

Drying occurs most rapidly at the end-grain through its very high permeability. The shrinkage close to the end-grain may therefore occur whilst the rest of the piece of wood still has a high moisture content and retains its swollen dimensions. The stress between the shrunk end-grain and the swollen inner wood frequently results in splits on the end-grain surfaces, although the term splits is normally used to describe cracks passing right through the piece of wood. In contrast, checks are more shallow, occurring as end-checks on the end-grain or surface checks on the other faces, perhaps diagonal to the edge of the board if spiral grain is present. Splits and checks may close with rewetting but, although they may then be virtually invisible, the strength has been lost. Serious splits are sometimes described as shakes, although this term is more correctly used to describe defects inherent in a particular tree rather than those developing as a result of faulty seasoning or kiln-drying.

The stresses and strains resulting from the rapid drying of end-grain in advance of the rest of a piece of wood have already been described but an identical situation occurs in respect of the entire surface of the piece of wood which naturally dries more rapidly than the inner wood and thus shrinks in advance. This defect is known as case hardening. Shrinkage of the outer zones tends to result in the development of surface checks. If rupture does not occur it is possible for the outer zone to be tension-set so that subsequent sawing will release this tension and result in spring or cupping; if tension-set is suspected it can be relieved by steaming. The tension in the outer zone may alternatively compress the inner zone which is wetter and thus more plastic, sometimes squeezing water out of the end-grain. Further drying may then result in the subsequent shrinkage of the inner zone without associated shrinkage of the dry set or case hardened outer zone, giving internal ruptures known as honeycomb checks or hollow horning.

Seasoning defects can still occur, despite great care in applying the most suitable drying schedules. Case hardening is perhaps the most common defect in certain species such as beech. When case hardening is anticipated or it has actually occurred the kiln must be operated at a high temperature and high humidity in order to plasticize the outer layers of the wood so that the case hardening is relieved and the wood recovers its proper dimensions. This treatment is applied only for a very short period, sufficient to relieve the case hardening without significantly affecting the moisture content of the piece as a whole. High temperature is also used during kilning in order to ensure sterilization of hardwoods against Lyctid beetle attack. In this connection it is interesting to note that kiln-seasoned wood is then far more susceptible to subsequent Lyctid beetle attack than air-seasoned wood, apparently because it possesses a higher starch content, as explained earlier in this chapter.

Joinery failure

Although it is normal to dry wood to a moisture content equivalent to the average atmospheric relative humidity anticipated is use, it is common to encounter movement problems.

Faults such as gaps appearing between floor blocks or boards are due to the wood drying after installation, either through inadequate kilning or perhaps rewetting between kilning and installation. A door or drawer jammed in humid weather may be exceedingly slack in drier conditions. Frames which introduce an end-grain surface in contact with side-grain will inevitably result in cracking of any surface coating system. In other situations the cross-sectional movement may become apparent as warping. The obvious solution to all these problems is to use only wood with low movement but this is not always realistic. The alternative is to impregnate the wood with chemicals which induce stabilization, although processes of this type are also frequently unrealistic because of the difficulty in achieving complete penetration.

Preferential wetting

The only theoretical alternative is to enclose the wood within a protective film in order to stabilize the moisture content. Paint and varnish coatings will act in this way, provided they completely cover the wood and are not damaged in any way. Unfortunately, whilst these coatings give good protection against rainfall, they are unable to prevent moisture content changes resulting from slow seasonal fluctuations in atmospheric relative humidity. As a result the painted wood will shrink or swell with changes in relative humidity, causing the surface coating to fracture wherever a joint involves stable side-grain in contact with unstable end-grain. Rain is absorbed by capillarity into the crack, yet the remaining paint coating restricts evaporation so that the moisture content steadily increases until fungal decay occurs if the wood is non-durable. It is frequently suggested that preservation provides a simple solution to this problem by reliably preventing decay but this ignores the fact that water also damages the paint coating. Wood is a hygroscopic material, covered with hydroxyl groups which have a strong affinity for water so that penetrating water will tend to coat the wood elements, displacing paint and varnish coatings. This failure is known as preferential wetting and is responsible for blistering and peeling in paintwork as well as loss of transparency in varnishes.

2.4 Fire

It is well known that wood is combustible, and a natural reaction to a fire disaster in a building is to demand that only non-combustible materials be used in construction in the mistaken belief that this will ensure fire safety. The reputation of non-combustible structural materials originates from observations on the behaviour of, for example, traditional heavy masonry but the same excellent fire performance is not necessarily achieved by other non-combustible materials. Fire in a building usually initiates in and involves mainly the contents, and the contribution from wooden structural members is small and usually limited to the later stages of the fire due to their size and position within the structure. Even in a completely wood-framed house there is a total of only one third more wood than in traditional mixed construction. It can therefore be appreciated that the combustibility of structural members is of limited importance compared with ability to withstand fire originating in furnishings and other building contents.

Flaming and charring

Whilst wood is certainly combustible, ignitability and rate of combustion are particularly important in most situations. Combustion cannot occur in the absence of oxygen so that large-section wood members burn only slowly at their surface which then progressively chars. As the temperature rises the wood first releases volatile components which flame on the surface and the residue then chars as a further increase in temperature occurs. The thermal conductivity of wood is very low, only 0.4% of that of steel

or 0.05% of copper, and the same order of conductivity as for cork, gypsum plaster and other insulation materials. Indeed balsa, a very low density wood, is used as an insulation material. The natural thermal insulating properties of wood therefore limit the rate at which heat can be transferred inwards from a burning external surface, and these insulating properties improve steadily as moisture is lost from the wood and charring progresses; charcoal is an even better insulator with a conductivity of only 30 to 50% of that of normal wood. Eventually the heat transfer from the surface is insufficient to release volatile components from the interior and the surface flaming then ceases. The charring rate also progressively slows down, unless heat is contributed from surrounding burning materials; the significance of the contents of the structure is again apparent.

Fire resistance

The charring rate is loss of dimension, whilst the burn-through rate is loss of weight. In large structural members the charring rate is important as the rate of change of dimension naturally controls the ability of the wooden component to continue to support the structural load. The charring rate varies with wood properties such as thermal conductivity and density which must therefore be considered when constructing a wooden barrier to achieve fire resistance, although design is of even greater significance. For example, there must be no gaps around doors or windows through which the fire can penetrate, or thin areas where early penetration can occur. Despite its combustibility wood possesses excellent fire resistance, largely because of its low thermal conductivity, and this not greatly affected by any form of treatment.

Ignition

The ignition point, usually about 270°C (518°F) for wood, is the temperature at which the rate of heating exceeds the rate of supply of heat and the fire becomes self-supporting with perhaps visible flame or glow. There is usually no ignition, even on the superficial surface, if the moisture content of deeper areas of wood remains above about 15%. A pilot source of flame is necessary in order to initiate ignition as spontaneous ignition of wood is possible only at exceptionally high temperatures. Large pieces of wood do not ignite easily, and ignition of smaller pieces is achieved only because of the rapid rate at which they reach the ignition temperature.

Flame spread

Flame spread across the surface of a piece of wood is really a series of ignitions, one area on fire acting as pilot ignition for the adjoining area. Flame spread is influenced by moisture content as in normal ignition but it also depends on the density and chemical nature of the particular wood. Chemical treatments can achieve considerable success in resisting ignition and thus preventing flame spread. Indeed, the propagation of a fire by flame spread is the most serious criticism of the use of wood as a structural material, yet this is the one property that can be readily modified by comparatively inexpensive treatment.

Smoke

Smoke generation represents the most serious hazard to human life during fires in buildings. The most harmful smokes arise from the plastics and synthetic fibres which are often used in furnishings, whilst the smoke from wood is comparatively innocuous. It is thus important to ensure that dangerous smokes are not generated by chemical treatments applied to wood to limit flame spread and propagation.

Despite its combustibility, wood thus possess distinct advantages when used as a structural material. The contribution of the structural woodwork to a fire is minimal, at least in the early stages, and wood has excellent resistance to fire penetration and suffers neither significant distortion nor rapid loss of strength. In comparison steel collapses when it reaches its yield temperature and concrete shatters or spalls, particularly if it encloses a steel frame. Where reinforcement bars are present in concrete they may be stressed at high tension and a rise in temperature will result in yielding with distortion of the structure even if complete collapse is avoided. Even if fire is controlled before the yield temperature of steel is reached there is still a danger that excessive expansion will rupture the building structure.

Preservation systems

3.1 Preservation mechanisms

The science, art and technology of preservation are concerned with the design, development and adoption of systems for preventing the various forms of wood deterioration. Decay by the wood-destroying Basidiomycete fungi is certainly the most important form of deterioration as it is inevitable whenever wood is exposed to dampness. It is true that some species of wood are more durable than others but none are completely resistant to decay, and a species that is said to be naturally durable will simply decay at a slower rate than one that is said to be non-durable.

Brown and White rots

There are three important factors when considering decay caused by Basidiomycetes. The damage caused by these fungi can be divided into two distinct types known as Brown and White rots respectively. A Brown rot generally destroys the cellulosic skeleton of the wood, leaving the lignin largely intact so that the wood becomes brown in colour and generally rather friable, perhaps developing longitudinal and cross-grain cracking. A White rot decays both the cellulose and lignin so that the colour remains virtually unaltered and the wood becomes soft and perhaps linty.

Fungus spread

A second important feature is the development of rhizomorphs by some fungi. These are composed of hyphae modified into strands which are able to conduct water and food from one part of the fungus to another. Some fungi are able in this way to transport moisture absorbed by one part of the fungus to another part where it can be used, for example, to condition wood prior to decay. In a similar way, nourishment or energy can be transported from an area of active wood decay to other parts of the fungus which are attempting to spread in a zone lacking nourishment. This might involve spread over masonry which may, of course, be a source of moisture but which can never be a source of nourishment. Alternatively the fungus may be attempting to spread into preservative-treated wood. In the case of some highly fixed preservative treatments the fungus may spread through the treated zone without causing damage but also without being affected itself, so that there is a danger that it may be able to spread to deeper zones which are unprotected through limited penetration of the preservative treatment. In other cases the availability of energy enables the fungus to actively detoxify preservatives, particularly certain toxic metals such as copper which can be de-toxified by the formation of oxalate. Highly developed rhizomorphs are a feature of only a few Basidiomycetes, particularly Brown rots, the best known being the Dry rot fungus *Serpula lacrymans*.

Fungus requirements

The third factor of importance concerns the

essential needs of a fungus for spore germination and development. The first requirement is for a source of nourishment, usually wood, but it must be appreciated that fungi are also capable of growing on a variety of other cellulosic materials. The second requirement is for water, most Basidiomycetes requiring an adequate moisture content within the wood that they are attempting to decay, as well as perhaps a sufficiently high atmospheric relative humidity. Indeed Basidiomycete attack is often invisible simply because the fungus avoids attacking the external surface of the wood where it is likely to be exposed to the dehydrating effect of an atmosphere of low relative humidity. The third requirement is for oxygen and this clearly operates in the opposite sense, tending to inhibit fungal activity deep within large pieces of wood. The fourth requirement is the need for a suitable temperature; low temperatures generally inhibit development whilst high temperatures may initially encourage the development of sporophores or fruiting bodies but eventually lead to the death of the fungus. If the temperature increase is slow the sporophores may be able to develop to a sufficient extent to produce spores which will be able to resist a further temperature increase. This will permit the fungus to redevelop when suitable conditions return, even if the original growth has been killed by the high temperature.

Preservation systems

Preservation systems rely on the elimination of one of these essential requirements in order to prevent fungal development. Chemical preservation treatment is essentially the elimination of a source of nourishment and will be discussed in detail later. The elimination of oxygen is a system that is actually used in practice where wood is waterlogged; for example elm is particularly susceptible to decay in damp conditions, yet it is widely used in marine and river defence works and even in boat planking where it gives excellent service provided that it remains saturated with water. Wood can survive for many centuries in marshy areas in this way, although unfortunately there are other chemical changes that slowly take place so that the recovery of archaeological wood presents special problems.

Structural design

The most widely used preservation system is to ensure that wood remains dry by taking appropriate structural precautions. Thus buildings are designed with roofs to protect the structure from rainfall. Projecting eaves, gutters and fall pipes are all devices to ensure that rainwater is dispersed clear of the structure. Walls are designed to resist penetrating rain, perhaps through cavity construction. Rainwater is still absorbed by the outer layers of walls constructed from porous materials and damp-proof membranes, flashings and soakers are provided to ensure that this dampness cannot penetrate to the interior. Damp-proof membranes are also provided to isolate the walls, joists and floors from dampness rising by capillarity.

These precautions are adopted primarily to ensure that the interior of the building remains visibly dry but it is essential to extend the principles to ensure that wood components remain dry. The main danger areas are wood in contact with porous external brickwork or masonry such as window and door frames and roof structures supported on outer walls. These precautions are not sufficient in themselves and it is generally also necessary to ventilate dead spaces under floors, in wall cavities and roof spaces, particularly flat roofs and frame walls. Unfortunately the Building Regulations in the United Kingdom usually limit ventilation in order to reduce heat losses so that condensation and wood decay problems frequently occur; these are discussed in more detail in the books *Remedial Treatment of Buildings* and *Defects and Deterioration in Buildings* by the present author. It must not be imagined that these problems arise solely through misconceived regulations, as one of the most common causes of condensation under floors is not limited ventilation but the use of air-conditioning equipment within the living space so that floor traffic surfaces are kept cool during warm weather and condensation is encouraged beneath, particularly in tropical areas, causing Wet and Dry rot problems similar to those encountered in buildings in temperate regions.

Decorative coatings

The simple precautions which have been described are usually sufficient to prevent decay in the carcassing or framing components of buildings but decay may still occur in wood which is exposed directly to the weather or which is in contact with porous brickwork or masonry exposed to the weather. Reference has already been made to the difficulties encountered in practice in ensuring that window and door frames are isolated from surrounding porous materials by proper damp-proof membranes.

As far as rain is concerned it is normal to rely on paint or varnish to protect exposed wood surfaces. Paint manufacturers often claim that paint functions as a wood preservative but fluctuations in atmospheric temperature, particularly differences in temperature between the interior and exterior of a building, result in redistribution of the water within wood window and door frames by evaporation and condensation so that this water becomes concentrated immediately beneath the cooler paint or varnish coat, usually the external surface during winter weather. Water penetration through a small damaged area of paint or absorbed through contact between an inadequately painted surface and adjacent porous brickwork or masonry will contribute to the moisture within the wood until eventually the moisture content reaches a level at which a fungal attack can develop.

Whilst it is true that an intact paint film prevents rain penetration, it is equally true that paint traps moisture within the wood. Random surveys carried out in England have shown that after a few years most window frames have moisture contents of perhaps 20 or 30% around the joints with the sills, often causing decay. This problem is usually due to the failure of the paint at joints where it is unable to tolerate the differential movement between side-grain and end-grain. In this way a crack develops which allows water to penetrate but the remaining paint inhibits evaporation so that moisture accumulates and decay ultimately occurs. Coating systems may thus actively encourage decay and in some countries it has become normal practice in recent years to require all window and door frames to be adequately preserved. Although this prevents deterioration of the wood it is not the complete answer to the problem; water accumulations still occur and therefore damage the adhesion between the surface coating and the wood. This problem is known as preferential wetting failure and will be discussed later in this chapter.

Preservation by isolation

The greatest value of paint may be to physically isolate wood from the attacking fungus as a painted wood component will appear to the fungus to be solid paint. Unfortunately the paint film can be penetrated by physical damage – the development of splits at joints, movement of the wood causing splits at edges, and the development within the wood of staining fungi which can bore outwards through the coating or alternatively settle on the coating and bore inwards. The various methods for avoiding these defects will be described later. In pursuing the theoretical preservation techniques of isolation it is apparent that these defects can be best avoided by impregnation rather than by superficial surface coating. Naturally it is also better if the impregnated material is toxic to the organisms that are likely to cause deterioration. The use of tar is perhaps the best example of the development of this type of treatment. Tar was originally applied as a superficial coating but later hot tar or impregnation treatments were adopted to encourage penetration and ultimately a special tar distillate, creosote, was selected for its toxicity towards fungi and its relatively low viscosity. Unfortunately all isolation treatments necessarily involve substantial alterations in the appearance of wood and the only alternative to the isolation principle is to adopt one of the various toxic systems.

Natural durability

Structural systems to ensure that wood remains dry represent the most important and most widely used preservation processes but there are situations where these cannot be used, such as where wood is exposed directly to the weather, ground contact or other hazardous conditions. It is possible to find woods which possess natural resistance to almost all biodeteriorating agencies; an indication of the durability of these woods which are most widely used is given in Appendix A. Generally, heartwood is much more durable than sapwood and those species with darker coloured and denser woods are usually most durable. Denser woods are less porous with more wood substance and less access for water and oxygen, while darker coloured woods often contain extracts which may be toxic to decay fungi or resins which may reduce water absorption. Unfortunately these general principles do not always apply. Thus resinous Scots pine or spruce is more durable than less resinous Silver fir, yet resinous Weymouth pine is not as durable as less resinous larch. Resin content is evidently not the entire story and certainly cannot account for the durability of most hardwoods.

The density principle is also frequently contradicted. Certainly the dry density of susceptible sapwood is invariably less than that of more durable heartwood, yet very low-density Western Red cedar is much more durable than some heavier softwoods. The cedar is darker due to the presence of extractives which are known to be toxic but it is clear that the natural durability

is not due to this alone but to a combination of numerous factors. Unfortunately durability is not always consistent within a species, although it is sometimes possible to judge probable durability just by colour and density. For example, in Scots pine the best durability is associated with slow growth and also with greater density and darker colour.

In all cases natural durability is a measure of resistance to decay; there are no woods that are completely durable in all circumstances as it is clearly necessary that nature should be able to dispose of dead trees, but the selection of naturally durable wood is a technically realistic way to avoid an unacceptable rate of deterioration and thus ensures the life of a structure. For example, the heartwood of European oak is frequently used for fence posts and Western Red cedar is used for window frames, greenhouses and external cladding. Unfortunately the use of naturally durable wood is not always possible; even if supplies are available the non-durable sapwood cannot always be rejected. It is frequently more realistic and less expensive to select wood for its desirable physical or aesthetic properties and then apply preservation treatment in order to ensure its durability.

Toxic preservation

Creosote has already been described as a wood preservative, developed from the isolation principle where reliability is improved by impregnation and by the use of a material that is toxic to the attacking organism. All isolation techniques result in a fundamental change in the appearance of wood and may be aesthetically unacceptable or even dirty, limiting their uses. The obvious alternative is to abandon the isolation principle with high retentions of materials of low toxicity and adopt instead low retentions of compounds of very high toxicity. Although creosote possesses only limited toxicity its retentions can be reduced for many uses, achieving cheaper and cleaner treatments. Originally, creosote was only applied in impregnation plants by the full-cell system which is designed to achieve the maximum retention possible but for many purposes empty-cell systems are now adopted which, whilst achieving almost the same penetration, reduce retentions by perhaps 40%; impregnation systems are described later in this chapter and creosote treatments are described in detail in Chapter 4.

There are many different toxic preservation systems. In principle, a preservative can be fungistatic in the sense that it can prevent the fungus from attacking the wood but will not necessarily kill the fungus. In contrast, in remedial treatment wood preservation, where fungus may already be present, a preservative must have fungicidal properties in order to ensure eradication. Most preservative systems rely on a direct toxic fungicidal action and on absorption by the fungus of toxic materials in sufficient quantities to prove fatal. Whilst this absorption is taking place some minor damage may be caused to the wood and some of the preservative components may be removed into the dying fungus. An adequate reservoir of fungicide is therefore essential to provide the required treatment-life in situations where there is continuous fungal attack as in posts and poles in ground contact. Thus the treatment must be permanent with good resistance to losses by leaching, volatilization and oxidation, yet it must remain available to the attacking fungus. The simplest toxic system may involve a compound of low solubility which, if applied at high retentions, will give an adequate life even when exposed to limited leaching. At the same time the limited solubility will be sufficient to ensure that if the treated wood becomes damp an attacking fungus will encounter a toxic solution. The best example of this type of treatment is the Timbor borate system which is applied either by diffusion into wet greenwood or by conventional pressure impregnation into dry wood. Timbor is a highly soluble sodium borate but the sodium ions are progressively neutralized by atmospheric carbon dioxide to give a boric acid deposit of relatively low solubility. Treatments of this type are suitable for situations where severe leaching is unlikely, such as within buildings where the decay risk is generally associated with either rainwater or plumbing leaks which are rectified when they become apparent, or condensation where leaching rarely occurs.

Fixation

One way to achieve good resistance to leaching is to make use of a preservative deposit which is insoluble in water but soluble in the presence of a fungal enzyme. In some cases pH alone may be involved. For example, the enzymes exuded by Basidiomycete hyphae are usually acid and therefore require a preservative that is soluble in the presence of acids. However, this creates the problem that the preservative may also be made soluble if the treated wood is naturally acid or if it is in contact with acid ground-water. A further point of interest with regard to pH is that enzymes may function only in acid conditions, so that wood with a high pH may have excellent resistance to decay. This principle has been applied in the preservation of wood chip piles at pulp mills by treatment with sodium hydroxide, although the same system cannot be used as a general wood preservative because of the caustic danger and the colour changes that result. However, some fungicides such as the various phenols are distinctly more toxic when applied as high pH alkali metal phenates than when applied as phenols.

One further effect of pH is its influence on spore germination. The spores of most wood-destroying Basidiomycetes germinate most reliably in slightly acid conditions, ensuring that germination occurs only on a wood surface which is acid when it is damp. In some cases the acid conditions are virtually essential for germination as in the cases of the Dry rot fungus, *Serpula lacrymans*, and it can be suggested that this may be because the fungus wishes to ensure that its spores can germinate only on wood which has already been subject to earlier Wet rot attack which generates the required acidity. For example, a plumbing leak in a building may result in a steady high moisture content in wood which will permit the development of a Wet rot such as *Coniophora puteana* but this fungus will be inhibited if the leak is repaired and the moisture content decreases. As the activity of the Wet rot ceases the moisture content will become more suitable for the germination of the spores of Dry rot, and this germination will be distinctly encouraged by the acidity of the wood caused by the Wet rot attack. An interesting point about this progression of events is that it apparently ensures the development of Dry rot in drying conditions in which this fungus alone is able to survive; if ventilation is restricted Dry rot can maintain the atmospheric humidity at the optimum for growth by exuding moisture on the superficial hyphae.

Detoxification

A Basidiomycete hypha exudes an enzyme solution which attacks the surrounding wood and the solution of the wood components obtained in this way is then absorbed by the hypha for nourishment. The tip of the hypha grows progressively into a cavity that is formed in advance by this enzyme activity. A fungicidal preservative treatment may restrict the ability of the enzymes to decay the surrounding woods, as in the case of a high pH treatment, or alternatively the enzyme may dissolve a toxic deposit which is then absorbed into the hypha, perhaps resulting in the death of the fungus. However, if a fungus which is well established on untreated wood spreads into a treated area it may have sufficient energy to tolerate the absorption of the toxicant, perhaps by converting it to a nontoxic form. Thus copper may be detoxified by the formation of copper oxalate, a process that is often apparent on wood treated with copper naphthenate solution as the characteristic green colour disappears a short distance in advance of the visible spread of the fungal hyphae. Detoxification requires energy and slow decay still occurs, so that the preservative treatment is entirely wasted.

Test methods

Failure of a preservative treatment in this way is not necessarily an indication that the preservative is unsuitable but usually that it has been applied at an inadequate retention and that the fungus has been able to develop on untreated wood, perhaps exposed by cross-cutting, drilling or other wood-working, after treatment. Alternatively it is possible that moisture content changes have resulted in the development of splits or checks which have permitted fungus to penetrate through the treated zone; it is essential that all preservative treatments should penetrate to a sufficient extent to avoid this danger. These factors must be clearly appreciated in any attempt to test wood preservatives. Normal laboratory testing methods invariably involve the establishment of a toxic limit or the retention at which the fungus is just controlled. Spores have little spare energy and germination is thus readily prevented on a treated surface, resulting in a low toxic limit. A slightly higher toxic limit generally occurs if treated wood is exposed to fungus with a limited energy source as in the European test EN 113 (British Standard 6009) in which treated wood blocks are exposed to a fungal culture on malt agar. A higher toxic limit is established where the fungus is spreading from an untreated wooden block which provides a more generous energy reserve as in the American and Nordic tests where the treated block is placed on top of an untreated feeder block. It is obviously desirable to ensure maximum reliability for any preservative system and these feeder block tests must be preferred as other tests may give unrealistically low toxic limits. In fact preservative approval systems are generally based on prolonged stake trials in natural ground contact conditions, and laboratory tests are used only to assess the preservative activity of individual toxicants or a newly developed formulation in comparison with an established preservative which is known to perform reliably in actual service.

Whilst referring to test methods it is worth noting that it is established practice in many laboratories to assess new wood preservative fungicides by dispersing them in malt agar and exposing them to fungi. This method is completely unrealistic as many toxicants are substantive on wood and may be detoxified by this process of fixation. This affinity of some toxicants for wood is one method of fixation but there may be other complex chemical processes involved before fixation occurs and it is essential that preservative chemicals should always be assessed as a treatment on wood and never alone. It has already been explained that, in addition, the spread of a fungus from untreated to treated wood represents the most severe risk because of the ability of the fungus to use energy from the untreated wood to detoxify the toxicant. The initial hyphae from germinating spores do not have access to such energy reserves and are therefore more readily controlled. Some fungicides, particularly types developed for agricultural use, are extremely efficient as spore germination inhibitors but this does not necessarily mean that they are equally efficient in resisting the detoxification mechanisms that are available to a well-established growth.

Application systems

Some wood preservative systems rely on a relatively superficial treatment applied by brushing, spraying or dipping. Any treatment with limited penetration is likely to be relatively inefficient where there is a danger of splits developing through movement resulting from changes in moisture content, as described in Chapter 2. This danger is particularly acute where the treatment is exposed to rainfall, as water is then trapped at the base of a split where, as evapo-

ration is restricted, it can be absorbed into the adjacent unprotected wood and can create ideal conditions for both spore germination and fungal development. Even in the absence of splits superficial treatments can provide only limited protection against fungal attack as in many cases hyphae will be able to penetrate through the treated zone, sometimes causing decay of untreated wood beneath.

Repellants

This hyphal invasion can be prevented only by the use of preservative systems with a direct toxic action or a repellent action; a repellent fungicide inhibits growth at a distance because it reaches the fungal hyphae by volatilization or leaching. The most efficient systems for superficial treatments are certainly those which make use of fungicides which have good resistance to both leaching and volatilization but which become toxic in the presence of a fungal enzyme. Another important feature of efficient superficial treatments is the use of penetrating solvents, particularly non-polar organic solvents, for the treatment of dry wood. These basic principles are also applicable to remedial-treatment wood preservatives.

Persistence

Organic solvent preservation systems generally involve the deposition of the toxicant as the solvent disperses by volatilization. The toxicants that are usually applied in this way are selected for their solubility in the carrier solvent system and this generally means that they are relatively insoluble in water and thus resistant to losses by leaching, although they must possess some solubility in water or in fungal enzyme systems in order to function as fungicides. The volatility of organic compounds depends on several factors but molecular weight is the most important. Lower volatility is associated with higher molecular weight but also usually lower activity so that it is necessary to select compounds possessing optimum combinations of permanence and activity.

In many cases a series of organic compounds is available with progressively changing properties in terms of toxicity, volatility and water solubility. For example, the chlorination of phenol typically results in compounds containing from one to six chlorine atoms. Generally the toxicity increases with the degree of chlorination but with an effective optimum fungicidal activity at the pentachlorophenol. As increasing chlorination also reduces volatility and water solubility it will be apparent that the pentachlorphenol is therefore the preferred compound, achieving optimum properties of activity and permanence. In the chlorination of naphthalene the optimum fungicidal activity occurs at only the dichloronaphthalene which is distinctly volatile and possesses limited permanence, but stomach poison insecticidal activity is still apparent in the tetrachloronaphthalene which is much more permanent and this compound has therefore often been used as a persistent insecticide in organic-solvent wood preservative formulations.

In all cases the surface deposits may be lost relatively rapidly by volatilization but the rate of loss of the total preservative deposit steadily decreases as the deeper deposits disperse much more slowly. The rate of loss depends on the diffusion gradient between the deposit and the free atmosphere, the gradient being shallower for deposits volatilizing at increasing depths, so that the permanence of a volatile treatment depends directly on penetration.

Fixation

These comments apply to simple deposits of organic compounds within treated wood but some organic systems involve more complex fixation. For example, copper naphthenate treatment suffers for an extended period from the characteristic odour of the appreciably volatile naphthenic acid which is liberated by slow hydrolysis. At the same time the copper becomes fixed within the cell walls and completely resistant to leaching, although it can be made soluble and detoxified by fungal enzymes. One practical problem with a system of this type involves evaluation. In short-term tests the naphthenic acid remains and contributes substantially to the activity but in long-term service the naphthenic acid is lost and preservation is reliable only if there is an adequate retention of copper. This is certainly the reason why many older copper naphthenate treatments proved inefficient and were actively detoxified by fungal growth; the retentions had been calculated on the basis of short-term results, influenced by the naphthenic acid residue and not on the basis of long-term service trials.

Chemical modification

The copper radical is a cation but there are many other cations which can preserve the cellulose structure in cell walls. Zinc is often used, particularly where a colourless preservative is preferred in place of a green copper product. Another cation that is now widely used is tri-*n*-butyltin, a highly active radical which fixes strongly to cellulose and is thus very resistant to leaching. In fact fixation is so good that some fungal hyphae can grow through the treated zone. This is not achieved through detoxification of the tri-*n*-butyltin but through failure of the fungal enzymes to activate the system. The treated wood does not decay despite the absence of toxicity and it is clear that the tri-*n*-butyltin has acted in a completely different way, modifying the substrate so that it has become resistant to the fungal enzyme system. Studies of tri-*n*-butyltin systems suggest that two groups are required for each cellulose chain, and at this ratio the cellulose is resistant to decay but hyphal invasion is able to occur. Unfortunately the lignin components in the wood remain unprotected so that some White rot fungi are still able to cause limited decay. One way to prevent hyphal invasion and White rot decay is to use higher retentions of tri-*n*-

butyltin but another alternative is to use a preservative formulation containing an additional toxicant such as a phenol which is particularly efficient in protecting the lignin.

Throughout this discussion, detailed reference has been made only to the tri-*n*-butyltin cation. In fact there are a variety of alkyl- and aryltin cations available but, as is explained in Chapter 4, tri-butyltin represents the optimum cation. These organotin compounds are available as halides, acetates naphthenates, etc. but the choice of anion is unimportant unless it is toxic in itself as all these compounds tend to hydrolyse to the oxide, so that tri-*n*-butyltin oxide is generally the most economic way to introduce this cation into a preservative formulation.

It is interesting to note that tri-*n*-butyltin oxide was first considered as a wood preservative as a result of fungicide tests in the laboratory, yet it is detoxified in the presence of wood and performs as a wood preservative only through its ability to modify cellulose, unless excessive loadings are used to ensure that unreacted compound is present. This clearly illustrates the need to include wood in any laboratory assessment of new compounds. It will also be appreciated that wood preservatives are not necessarily toxic and a candidate compound need not be rejected simply because it does not possess fungicidal activity. Acetylation has been considered as a method for stabilizing wood and it has been found that it also functions as a preservative, but it is not a toxic system and will function reliably only on woods that are sufficiently permeable to be impregnated throughout their thickness.

Toxic precipitates

Aqueous preservative systems must possess a fixation mechanism as there is otherwise a danger of loss through leaching whenever damp conditions occur. Indeed, an aqueous preservative which lacks a fixation system is entirely unsuitable for use against Basidiomycete fungi

which represent a risk only when the conditions are damp and when there is therefore a danger of leaching. Aqueous preservatives can achieve fixation in several different ways but many have been designed, at least initially, to give insoluble precipitates within wood. Copper can be solubilized in ammonium solution and subsequently precipitated through loss of ammonia. Sodium pentachlorophenate is precipitated as pentachlorophenol through the neutralization of the sodium ion by absorption of carbon dioxide from the atmosphere. In some other multisalt formulations precipitation is achieved by a double decomposition, typically by applying a toxic solution such as copper sulphate followed by a fixation solution such as sodium chromate to give a precipitate of insoluble copper chromate. Although this is apparently a realistic fixation system it does not actually perform in this way within wood and, whilst it is found that sodium chromate precipitate is present, it is also apparent that the copper and chromium are fixed independently to the wood elements. Indeed it has been shown that the application of copper sulphate solution alone will achieve limited fixation of copper in this way.

Two-stage treatments are commercially unrealistic as they involve double treatment costs so that modern multisalt preservative systems involve single treatments, relying on loss of a volatile component or the influence of the wood elements to achieve the required fixation; this will be described in greater detail in the discussion of the fixation of acid copper–chromium and copper–chromium–arsenic preservatives in Chapter 4. One interesting observation is that a copper cation appears to preserve the cellulose whilst a chromate anion appears to preserve the lignin, a system exactly parallel to that described in connection with organic-solvent preservatives where copper and tri-*n*-butyltin cations were described as cellulose preservatives and phenol anions were described as lignin preservatives.

Resistant fungi

It must be emphasized that fixation systems must achieve resistance only to losses by volatilization, leaching and oxidation, leaving the preservative accessible to fungal enzyme systems or alternatively modifying the substrate so that it is unaffected by the fungal enzymes. In the case of toxic systems the sensitivity of the Basidiomycetes varies greatly. Presumably the fungi are likely to be most sensitive to the toxicants that are most readily solubilized by their enzyme systems and most resistant to those that they are most readily able to detoxify, perhaps by oxalate formation. For example, *Coniophora puteana* is resistant to many preservative toxicants and, as it also occurs widely and causes substantial damage, it should always be included in laboratory tests to evaluate preservatives. *Lentinus lepideus* is known to be resistant to tar-oils whilst *Lenzites trabea* is resistant to arsenic and *Poria* species to copper; in initial laboratory evaluation of preservatives containing these toxic components the appropriate resistant fungi should always be employed.

Soft rot tolerance

The Soft rotting Ascomycetes and fungi Imperfectii are generally controlled in the same way as the Basidiomycetes but their distinctly different behaviour causes some unusual difficulties. The Soft rots were first discovered in Britain as a result of extensive investigations into the failure of wood fill in water cooling towers. This fill consisted of softwood impregnated with Wolman (FCAP) salt-type preservatives but deterioration had still occurred under the warm saturated conditions and the cause could not be identified. The decay commenced on the external surface of the slats, producing the surface softening that is characteristic of this form of decay. There was a complete absence of superficial visible fungal growth but eventually hyphae were identified within the cell walls. As the hyphae were visible only under the microscope the organisms were at first described as micro-fungi.

It was soon established that these Soft rotting fungi were resistant to several types of preservatives, particularly the fluorine types widely used at that time. It was also observed that cooling tower fill treated with copper–chromium preservatives was generally free from serious Soft rot attack, and the ultimate solution to the cooling tower problem was to treat all softwood fill with copper–chromium or preferably copper–chromium–arsenic preservatives, the arsenic content improving preservation against the copper-resistant *Poria* species that are encountered in the drier components of the towers, such as structural supports and mist eliminator slats.

It has since been observed that ground-line attack in poles is often largely caused by Soft rot fungi. Whilst copper–chromium–arsenic preservatives are reasonably efficient in preventing Soft rots in softwoods they are not so efficient in hardwoods, a serious problem in Australia where *Eucalyptus* poles are extensively used. Damage is particularly severe in tropical areas and can be attributed largely to the fact that although the copper–chromium–arsenic preservative has been distributed on all accessible surfaces within the wood, it has not deeply penetrated the cell wall so that the Soft rot micro-fungi are able to explore within the cell wall without being affected by the preservative treatment. This micro-distribution problem is a matter of very serious concern and the subject of extensive current research.

Staining fungi

The control of staining fungi presents entirely different problems. These fungi invariably develop on the sugar and starch cell contents without affecting the cell walls or the structural strength of the wood. A preservative that fixes to the cell wall is therefore completely ineffective against stain fungi which respond only to preservatives which can exert an influence over the cell contents.

Stain in greenwood

There are two separate stain-control problems. The first problem concerns the development of stain in freshly felled greenwood possessing a high moisture content. It is frequently suggested that stain can be avoided by rapid kiln-drying but the rate of moisture content reduction is often too slow and the elevated temperatures frequently encourage heavy stain development before drying eventually achieves control. If drying is sufficiently rapid to prevent stain development the cells are killed by the elevated temperatures before they have been able to utilize the sugar and starch cell contents, so that there is then a distinct danger that stain will redevelop if the wood subsequently becomes damp.

Some stain-control treatments rely on spore germination control or direct toxicity to hyphal invasion. Pentachlorophenol is one of the toxicants that has been most widely used in stain-control treatments; it has the advantage of being slightly soluble in water and also slightly volatile so that it is able to diffuse to a limited extent and achieve some control beyond the limit of physical penetration. This ability to exert a distant action is particularly important as stain control treatments are generally superficial dip or spray applications as more deeply penetrating treatments cannot be justified as only temporary protection is required.

Freshly felled wood has a very high moisture content and it is therefore normal to apply the treatments in the form of an aqueous solution, so that pentachlorophenol is generally used as sodium pentachlorophenate. The sodium ion is rapidly neutralized by the natural acidity of the wood and carbon dioxide in the atmosphere but it still ensures an increase in the pH of the treated wood which enhances the activity of the pentachlorophenol. Indeed, the addition of

excess sodium hydroxide, sodium carbonate or sodium tetraborate ensures the maintenance of a high pH and much greater efficacy, so that a reliable treatment can be achieved at far lower pentachlorophenol retentions. The most efficient treatment is probably a combination of sodium pentachlorophenate and sodium tetraborate, the borate providing efficient pH control but also functioning as an additional toxicant and broadening the spectrum of activity of the system.

Borate gives excellent control of many staining fungi but unfortunately allows other fungi and particularly surface moulds to develop if it is used alone. The addition of a limited amount of pentachlorophenol is sufficient to avoid these problems; a system containing only about 20% of the normal pentachlorophenol content provides a safer, less expensive and more reliable treatment than sodium pentachlorophenate alone. Organomercury compounds have also been used for stain control but they have limited persistence.

Recent developments have included the use of dispersions of insoluble fungicides such as Captafol and Benomyl, but these are generally far less efficient than the sodium pentachlorophenate systems, apparently because their activity is confined to the superficial surface, a characteristic which is generally unacceptable as a stain-control system must prevent the development of internal stain and not simply keep the surface clean. One further recent development is the addition to stain-control formulations of water repellents and other components such as polyethylene glycol in order to reduce the development of splits through the treated surface and improve the control of internal stain.

Bark-borers

Stain-control treatments are generally applied to sawnwood immediately after conversion but there is also a danger that stain may develop in logs if conversion is delayed. This is a serious problem in tropical areas where there is also a distinct risk of bark-borers, particularly the Scolytid and Platypodid Ambrosia beetles, which cause physical damage in addition to introducing stain fungi. Logs in tropical forests are therefore often treated immediately after felling with formulations containing both stain control fungicides and contact insecticides.

Stain in service

Stain may also develop in service through condensation beneath paint or varnish coatings. Problems with decorative systems will be described in detail later but it is appropriate at this point to consider the staining problem alone. The stain fungi are often the same mixed species that are responsible for stain in freshly felled greenwood, although the same species do not necessarily dominate.

Aureobasidium pullulans is a particularly important stain in service. It may develop in wood through condensation under surface coatings but it then attacks the surface coating medium, eventually causing small unsightly black pustules to form on the coating surface. The formation of these pustules is associated with the development of a hole through the coating which allows rainwater to penetrate and thus to accumulate beneath the coating. This species is also able to develop on the surface of the coating, boring through it into the wood beneath.

This is a relatively new problem, at least in decorative paint systems in the British Isles, apparently because it has been usual in the past for primers and undercoats to contain lead pigments which were sufficiently toxic to control these fungi. The problem is particularly acute in countries such as Germany, where it is normal to prime wood with dilute linseed oil before applying a coating sustem, thus directly encouraging the development of *Aureobasidium pullulans*. Attempts have been made to treat wood and also to add toxicants to priming oil

systems but generally these precautions have been relatively ineffective, apparently due to their limited persistence. It is probable that toxicants or toxic pigments must also be added to the paint system if comprehensive control of stain development is to be achieved.

Natural control

There have been several attempts to achieve natural control of staining and wood-destroying fungi. Some bacteria and mould fungi which are themselves virtually harmless can inhibit the growth of fungi. The most efficient system of natural control developed so far involves the inoculation of freshly felled green sawnwood with the mould *Trichoderma*. Unfortunately this mould growth is very unsightly but it is superficial and is removed if sawnwood is subsequently machined. At this stage it cannot be envisaged that a naturally antagonistic control system will be developed which will be commercially realistic, although structurally harmless bacteria are already used as a means for increasing the permeability of spruce sapwood which is usually resistant to preservative penetration, enabling it to accept pressure-impregnation preservative treatments.

Borer control

Preservation of wood against wood-boring insect attack is similar in principle to preservation against wood-decaying fungi but there are several practical differences. Insects are, of course, far larger than the exploring hyphae of a fungus. In addition, an insect is capable of movement and thanks to this ability and its various sensory organs it is able to select the wood that it will attack. With a fungus the situation is quite different; spores are dispersed at random and the developing fungus must then attempt to infect whatever substrate happens to be available. Despite this fundamental difference between the two types of attack the principle of isolation is certainly one of the most important techniques of preservation in both cases. Surface coatings are perhaps more effective against insect attack than against fungi as the development of cracks does not generally permit wood-borer infestation to become established, unless the female insect is able to lay eggs through the defect in the coating and larval development can subsequently take place in the wood beneath. This is unlikely as there is no reason to suppose that an insect can understand that a coating is concealing wood which might be an attractive site for egg-laying, and this slight danger is further reduced if an impregnation treatment is used instead of a superficial coating.

Termite shields

Fungal spores are dispersed widely in the atmosphere and similarly wide-spread damage can be created by flying insects, but many wood-borers are wingless. The most important wingless group is probably the subterranean termites which can be prevented from infesting building wood by the provision of physical shields which they are unable to negotiate; these shields are described in Chapter 5. Termite shields suffer from the serious disadvantage that a slight error such as subsequently installed plumbing or wiring can render them entirely useless. For this reason it is more usual to prevent subterranean termite attack in buildings by soil poisoning. Although this technique involves the use of toxicants the actual system is still a shield involving the isolation of structural woodwork from the source of infestation by poisoning the soil with an insecticide possessing a repellent action.

Repellants

With fungi, a repellent action invariably involves the volatilization or solution of the toxicant to a sufficient extent to affect growth and thus inhibit the spread of the fungus towards the treated area, but insects have senses and are able to move away if they dislike the conditions. Generally, repellants affect the olfactory senses but they do not necessarily involve toxicity. About 200 years ago it was suggested that snuff was such an unpleasant material that it would repel any type of borer, although the results of actual tests against marine borers showed that it had no action whatsoever.

Repellency is not confined to the olfactory senses. For example, some species of termite will not explore or even build galleries across white surfaces, and there are many other reports of insects which will not lay eggs on surfaces of a particular colour. It is also known that female insects frequently taste wood before laying eggs in order to ensure that the substrate will be suitable for the emerging larvae. In the same way insects which bore in the adult stages may also be discouraged from causing damage, even without the use of toxicants. Natural durability against insect attack for many tropical woods can be attributed to their high silica content, and this observation has resulted in the simplest method for protecting polyvinyl chloride (PVC) wiring against termite attack as the addition of silica results in greatly extended life.

The concept of using silica to prevent borer attack is not new; it was suggested as early as 1862 that silica deposits should be used to improve resistance to marine borer attack. The idea was revived in about 1950 with little success, although this may be due to the failure to deposit the silica in the correct form. It is probable that silica acts as an irritant or abrasive, affecting the mouth parts of the borer, but in a memorable debate at a British Wood Preserving Association Convention in Cambridge in 1959 it was suggested that boron compounds might also be non-toxic in their action, perhaps a 'sharp boric acid crystal up the ovipositor' being the true explanation for the excellent resistance to egg-laying with borate treatments! This suggestion may appear to be rather amusing but wood-borer insecticide development has generated even stranger suggestions. Persons who answered an advertisement publicizing a 'completely reliable method for killing woodworm' were sent a small box containing two blocks of wood and a sheet of instructions which told them to 'place the woodworm on the first block and hit it with the second block'!

Insect traps

A more serious suggestion involved the exposure of blocks of alder, a wood species that is very susceptible to Common Furniture beetle attack. The intention was to distract egg-laying females from furniture or structural woodwork, burning the blocks annually to prevent the development of a new generation of adult insects. The destruction of the blocks was often forgotten and their presence actually encouraged the infestation.

The sex life of adult wood-borers has also attracted attention. With flying insects it is possible to release hormone from a trap which attracts insects of the opposite sex which can then be electrocuted or controlled by contact with a toxic deposit. A more efficient technique is to sterilize the attracted insects which can then be released to ensure sterile matings as most female beetles mate only once. Another serious proposal for an insecticide is based on the observation that kidney tubule function is stimulated by a particular hormone so that its use may cause the death of the insect through dehydration, but this is really a direct toxic insecticide.

It is obvious that wood-borers require wood but it must be of a suitable species, an appropriate part of the tree and in the right condition. Removal of bark will be sufficient to prevent attack by borers that are dependent on bark such as wasps, longhorn beetles, Bostrychids and other bark-borers, such as *Ernobius mollis*, and rapid drying after bark removal will also inhibit Ambrosia beetles; details of these and other borers are given in Appendix B. The removal of the bark is not sufficient to prevent damage by some borers such as the Lyctids,

which generally attack any large-pored wood containing adequate starch. Even if sapwood is not susceptible to Lyctids it is usually susceptible to other borers such as the Anobids but it is entirely unrealistic today to remove all sapwood in an attempt to avoid this danger and the only alternative is the use of toxicants to provide protection.

Toxic preservation

There is no application in wood preservation for insecticides which must be applied topically in order to achieve control and it must be generally assumed that the treatment must achieve control through contact with treated wood. With contact insecticides, control may be achieved through simple contact as their name implies. A larva boring into treated wood will tend to absorb the insecticide wherever the body is in contact with treated wood but the situation is rather different with an adult beetle which has a hard protective covering and which may only absorb the insecticide through the mouth or claws. It is not sufficient to develop an insecticide as it may also be necessary to provide an access system to the insect, perhaps involving a special oil which will enable it to be absorbed through the insect cuticle or claws, or which will encourage tasting.

In contrast, the stomach insecticides must be ingested by the insects while boring so that some damage will occur before the insect dies, particularly when an indirect-action toxicant is involved. Many insects do not possess enzymes which can convert wood to an absorbable form and rely instead on intestinal symbionts such as bacteria or yeasts. A preservative may act by controlling these symbionts, causing the insect to die of starvation, but it will continue to eat and may cause significant damage before death. Many insects are also encouraged by or dependent on fungal attack. With some borers such as wood weevils damage can be prevented by ensuring that fungal decay is unable to develop. Some termites attack dry wood, removing it to their nest where they convert it to food in fungal gardens; wood treated with a fungicide will eventually destroy the fungal gardens and control the termite colony but significant damage may occur before control is achieved. Even direct-action stomach poisons can act in many different ways, perhaps affecting the neural system or alternatively affecting cell metabolism either by retardation or acceleration. The availability of adequate energy will permit detoxification as with fungi, particularly in larger insects where the available stored energy may be considerable relative to the absorption of toxicant from preserved wood, and it is essential to use adequate retentions of preservatives in order to achieve reliable control.

It is relatively easy to protect freshly felled non-infested wood. Any treatment will tend to discourage a female from laying eggs, but egg larvae are small with limited reserves of energy so that they are very sensitive to toxicants. It is therefore best to evaluate preservative treatments using larger established larvae which are likely to be more resistant.

Eradicant insecticides

The eradication of an established infestation is not easy. It is difficult to ensure that the treatment penetrates sufficiently deeply to be lethal to larger larvae. Indeed, many eradication treatments rely simply on preventing the ultimate emergence of adult beetles which, coupled with the prevention of egg-laying, ensures that the infestation will eventually be eradicated, although the boring within the wood may be appreciably extended before this is achieved.

One particular difficulty with insecticidal preservatives is to achieve adequate permanence. Fixation invariably means a loss of repellent or contact action so that insecticidal preservatives with long life generally rely solely on a stomach-poisoning action so that wood must be ingested before the intestinal enzyme systems release the toxicant, unfortunately allowing some tasting damage before the borer is eventually killed. In

most cases this tasting damage is insignificant, representing a problem only when the danger is from very large numbers of adult insects as with termite attack in tropical areas and gribble attack marine conditions. In many cases the eradication of the borer infestation is more important than preservation against further attack, particularly with Lyctid beetles which attack the sapwood of a limited number of hardwoods but only within a short time after felling and conversion. Heat treatments, sometimes forming part of a kiln-drying cycle, have been widely used to achieve eradication although they give no protection and there is a normal danger of reinfestation.

Heat and fumigation treatments have also been used against Wood wasps with much, greater success; the wasps lay eggs only through the bark, so that reinfestation of sawnwood is impossible. The Australian quarantine regulations were originally introduced to prevent wasp larvae in imported sawnwood from emerging, mating and infesting valuable new conifer plantations, and both heat and fumigation treatments were approved as methods for ensuring that imported wood was free from infestation. While these are examples of eradication problems encountered with new wood, it will be appreciated that remedial *in-situ* preservation treatment is necessary where infestations have become established in, for example, buildings and boats; whilst the formulations typically used for these treatments are described in Chapter 4, the techniques involved are highly specialized and are described by the author in more detail in the book *Remedial Treatment of Buildings*.

Natural control

Even very active infestations of wood-boring insects have been known to die out naturally without the need for remedial preservation treatment. In many cases this control can be attributed to the action of parasites and predators – these are described more fully in Appendix B. The best known predator is *Korynetes coeruleus*, a handsome metallic-blue beetle that is often found in association with heavy infestations of the Death Watch beetle. Amongst the parasites, mites are most frequently found, particularly in Powder Post and Common Furniture beetle infestations, both of which also support minute ant-like parasites. Whilst these predators and parasites can eliminate infestations naturally, they cannot be realistically harnessed as a method of either preservation or even eradication. The only possible method of biological control that has yet been devised is based on the observation that the bacterium *Bacillus thuringiensis* is frequently found in association with termite colonies in decline, and it has been found that similar decline can be induced in otherwise healthy colonies when it is introduced. Bacteria are very resistant to inhospitable conditions and it is therefore possible to envisage a deposit of this bacterium on wood which would become active only through direct contact with a suitable host insect.

Dampness generally results in fungal decay in wood and this encourages the development of insect borer infestations; a general-purpose preservative should not be fungicidal alone but should also be insecticidal. In the same way an insecticidal preservative should preferably contain fungicides to reduce the danger that a fungal development will encourage unusual borer activity. There are few situations where a single-action preservative can be clearly justified, perhaps fungicidal properties alone being essential in wet mining conditions whilst an insecticide alone might be justified for the protection of furnishings. In the case of freshly felled green wood a fungicide is frequently applied to avoid sapstain damage, although in tropical conditions and insecticide is often added, as the stain infection is introduced by pinhole borers which carry spores of the Ambrosia fungus which infects the walls of the galleries and ultimately becomes the food for hatching larvae. Clearly this fungal infection is dependent on high mois-

ture content and protection is required only for the comparatively short period until the wood is too dry to support the staining fungi. In this type of situation a stomach insecticide is generally unsuitable as tasting damage can occur – before dying an insect may penetrate through the superficial protective treatment, allowing deep stain development to occur although the surface might appear to be entirely clean.

Marine preservation

Marine borer preservatives generally function in the same manner as those intended to give protection against wood-boring insects and decay fungi. Preservatives against crustacean borers operate in precisely the same way as those against insect borers, functioning as repellants, contact toxicants or stomach poisons. Indeed, one feature is the excellent activity of contact insecticides such as the organochlorine and pyrethroid compounds against crustacean borers, presumably because the insecta and crustacea are closely related, both being in the phylum Arthropoda.

Surprisingly little use is made of this observation and these contact insecticides are rarely used in commercial marine borer preservatives. This may be due to the fact that these toxicants are virtually inactive against the molluscan borers such as *Teredo* species which behave in an entirely different manner. The larvae of these borers are minute and capable of little movement so that they are distributed in the sea in a random manner similar to fungal spores in the atmosphere. If they settle on a suitable wood substrate they metamorphose and ultimately develop into a borer concealed within the wood, but it is not clear whether the borer derives much or even any nourishment from the wood which it removes to provide a gallery for its accommodation. The main requirement is for a preservative treatment which will prevent the metamorphosis of the settling larva and this is often achieved with preservatives that are considered to be predominantly fungicidal in action, such as creosote. In contrast, creosote gives little protection against the crustacean borers such as *Limnoria* species which are able to cause significant damage before they are affected so that eventually the treated layer crumbles away, perhaps exposing untreated wood beneath.

All toxic marine preservatives must be completely resistant to leaching to give reasonable life to treated wood, although with the exception of the contact insecticides this necessarily means that some tasting damage must occur before attacking crustacean borers are affected. In practice the tasting damage is kept to a minimum by using very high loadings of stomach insecticides which will ensure death before significant damage can occur. Non-toxic preservatives have also been considered, particularly silica deposits in wood, an idea that originates from the observation that many naturally durable woods possess high silica contents. Trials have not been entirely successful, as mentioned earlier in connection with preservative systems against insect borers, but it is probable that the silica must be deposited in a particular crystalline form if the desired protection is to be achieved.

Decorative coatings

Although they are generally applied for aesthetic reasons, decorative systems such as paint and varnish coatings are often claimed to have preservative value. It was explained earlier in this chapter that there are two points of weakness with coating systems. Minor imperfections due to discontinuous application, damage or very thin coatings on sharp edges, and inadequate paint applied on concealed surfaces in contact with damp brickwork and masonry, may permit limited absorption of water which is then trapped beneath the coating where it accumulates. There is a danger that fungal decay will eventually develop, although this is perhaps less likely than preferential wetting; wood is hydrophilic so it prefers to be covered with water rather than with a non-polar oil such as a normal coating system, so that accumulation of water beneath a coating results in loss of adhesion between the coating and the wood. Failure through preferential wetting normally becomes apparent first at joints and then spreads with the development of opacity under varnish systems. Failures of this type invariably occur and it is necessary to regularly strip the coating system to the bare wood and to apply a completely new system in order to maintain the appearance, a very expensive maintenance liability.

It is frequently claimed by paint and varnish manufacturers that these failures result from the use of a poor paint system or careless application. In Britain it is considered that a paint system should consist of an initial primer containing excess oil which will penetrate the pores, blocking them as it dries and thus 'killing the suction.' This primer is then followed by an undercoat with a very high pigment content which is intended to fill irregularities in the surface and to achieve opacity. The finish coat possesses a high varnish or binder content in order to give a hard durable surface. In the case of varnish coatings only a single composition is generally employed but it is often thinned with solvent for the initial priming coat. This is intended to achieve improved penetration of the pores to establish good adhesion. In fact the dilution with the solvent simply reduces the amount of varnish that is applied in the primer coat. Whilst the dilute varnish has a reduced viscosity this does not actually affect the viscosity of the varnish medium, which is controlled by its molecular size, so that the inactive diluent solvent tends to penetrate and leaves the varnish components on the surface, so that there is no practical advantage with thinning. In other European countries such as Germany a priming oil is often used in place of a pigmented primer. This priming oil consists of a solution of drying oil, essentially similar to the thinned varnish used in a British varnish coating system and subject to the same criticisms.

Although it is true that good quality paints and varnishes combined with careful workmanship achieve the best results, none of the normal systems is currently able to resist preferential wetting. This is particularly apparent under varnish, where it results in the development of dark stains and opacity. It is generally accepted that these failures are associated with the presence of water under the coating. Some paint manufacturers try to reduce the permeability of their coating systems to make them very resistant to liquid water penetration, but diffusion of water vapour with seasonal changes of atmospheric relative humidity is still able to occur. Unless the wood is unusually stable some swelling and shrinkage is inevitable with gain and loss of water vapour. This movement is virtually confined to the cross-sectional dimensions, the longitudinal dimensions remaining stable. Some paint manufacturers have attempted to produce more flexible coating systems able to withstand this cross-grain movement but the resulting systems are still unable to resist the shear stress where end-grain is in contact with side-grain at a frame joint. This stressing invariably cracks paint and varnish coating systems, allowing liquid water to penetrate by capillarity, although subsequent evaporation is prevented because most of the wood surface remains protected by the paint and the water simply accumulates, introducing a danger of decay and preferential wetting failure despite the conscientious application of an excellent paint system.

Some other paint manufacturers attribute the fault to condensation beneath the paint coating through the redistribution of moisture in the wood through daily thermal gradient changes. Some coatings are therefore designed to allow water accumulations to disperse by evaporation through the coating system. In fact any coating system that permits moisture vapour to disperse outwards also necessarily possesses poor

resistance to seasonal changes in atmospheric relative humidity and is usually particularly susceptible to stress cracking through movement.

Water repellants

Despite extensive efforts over many years the surface-coating industry has completely failed to solve the problem of coating external wood joinery (millwork) such as windows, doors and frames. In the United Kingdom and several other countries there are now several recommendations and specifications which require external joinery to be constructed from naturally durable wood or wood which is adequately preserved to avoid fungal decay, but preferential wetting failure remains a serious problem.

Water-repellent treatments have been proposed to prevent migration of water absorbed into joints damaged by movement cracking of the coating system. These treatments certainly delay failure but they seldom represent a complete solution to the problem as many of them, particularly those based on waxes and resins, are just as susceptible to preferential wetting failure as the paint systems which they are designed to protect. Indeed, many water-repellent formulations are essentially similar to the thinned varnish or priming oil systems that have already been described, and if these formulations are based on drying oils they can introduce a further problem as they may encourage the development of stain fungi under the paint or varnish system. Water introduced by any means, even condensation redistribution of the moisture within the wood, will eventually accumulate beneath the coating system when the weather is cold and stain fungi can then develop, causing the darkening that is characteristic of the progressive weathering of a varnish system.

The staining fungi are similar or identical to those that develop on freshly felled sawnwood but some species, particularly *Aureobasidium pullulans*, do not confine their activities to the wood but are also able to attack the coating medium. The inclusion of normal wood-preservative components in a water-repellent treatment is not sufficient to give resistance to these particular fungi which can cause the development of bore holes through the coating system, eventually resulting in black dust, actually sporophores, on the surface. These holes can permit further water penetration through the coating system, thus encouraging the ultimate development of wood-destroying Basidiomycete attack as well as preferential wetting failure. Even if wood is treated with a toxicant that is particularly active against these fungi and thus free from staining under the coating, the result may not be entirely satisfactory as spores can settle on the external surface and bore inwards.

In the British Isles these problems have always been common on varnish but were virtually unknown in the past on paint coatings, apparently because old lead primer and undercoat systems were able to resist these fungi. Lead paints are not widely used nowadays because of their toxicity but it is possible to achieve similar protection by the use of zinc pigments. It will be appreciated that pigments cannot be used in clear varnish systems but suitable zinc compounds are available, such as resinates. Even then the result may not be entirely satisfactory as stain may develop in the wood beneath the varnish.

There is really a need for an entirely new priming system which can perform a variety of functions. It must be a wood preservative with activity against both the wood-destroying Basidiomycete fungi and also the staining fungi and superficial moulds, perhaps involving the incorporation of a single toxicant with a very wide spectrum of activity or multiple toxicants. In addition, the primer should act as a water repellant to give protection against rainfall or other sources of moisture before the finishing coats are applied. It is not sufficient for the treatment to be water-repellent in the sense that it relies on a contact angle action alone but should perhaps be best described as a water-

proofer as it must also block the pores. This pore blocking will reduce changes in moisture content of the wood arising through fluctuations in atmospheric relative humidity, but pore blocking is also necessary to form a foundation on which to apply the undercoat system. Finally, the primer system should achieve permanent bonding to the wood substrate so that the entire coating system is resistant to damage by preferential wetting.

It has been suggested in the past that these various functions can best be achieved by incorporating toxicants into a conventional pigmented primer paint but this is entirely unrealistic; the loading of a primer paint on wood is very low and it is impossible to incorporate an adequate amount of carrier solvent in order to achieve the penetration and distribution of the preservative components. An alternative is to add the pigment and binding components in a primer to a normal organic solvent wood preservative, but penetration of the toxicants can be achieved only if the system possesses low viscosity and this necessarily results in difficulty in maintaining the pigments in suspension. Finally, neither of these systems achieves any control of preferential wetting.

A more realistic technique is to achieve the desired resistance to preferential wetting in a penetrating solvent system, perhaps applied by some advantageous technique such as a conventional vacuum/pressure or double vacuum system, and then to add other compatible components in order to achieve the additional desired functions. Some long-chain alcohols have been suggested as a means for achieving resistance to preferential wetting but, although they perform reasonably well in short-term laboratory trials, they slowly hydrolyse and separate from the wood and they possess no long-term advantages over conventional wax water repellants. Excellent results are achieved with certain organometallic compounds, including those based upon group IV/IVb such as silicon and tin. Organotin compounds such as tri-*n*-

butyltin oxide are used as preservatives but if they are required to give resistance to preferential wetting much higher loadings are necessary and some practical problems arise such as volatile losses which may introduce a toxic hazard. Quaternary ammonium compounds can also give good resistance to preferential wetting but the quantity required is critical; if excess compound is applied, dry adhesion of subsequent paint films is seriously affected.

Clearly the development of advantageous priming treatments of these types must be a proprietary development but, although such processes were first proposed in 1968, there have been no realistic commercial developments since and it must be assumed that the coating industry has little interest in achieving permanent paint coatings, perhaps because most of its market arises through the maintenance of defective paintwork! It must not be supposed that these priming systems can function only under pigmented paint as pore blocking can be achieved using resins, a universal system which can be used under both paint and varnish coatings. If these systems can successfully overcome preferential wetting so that the coating system permanently adheres to the wood, maintenance becomes simply a matter of washing or sanding the surface and applying a further gloss coat to maintain the appearance. As labour costs increase throughout the world it becomes more important to produce durable systems than to rely on regular maintenance to achieve reliability.

Decorative preservatives

Surface coatings such as paint and varnish are not essential for the decorative use of wood. Bare wood can be very attractive and perfectly satisfactory in service provided species is selected which possesses natural resistance to decay and low movement to give resistance to distortion and splitting. Alternatively non-durable species can be used if appropriate preservative treatment is applied to achieve these

requirements. However, the wood will suffer from loss of colour, principally through leaching and oxidation of extractives and the development of surface-staining fungi, these factors generally combining to give the grey shades that develop when wood is naturally weathered. This natural colouration tends to be patchy related to exposure to rainfall with protected areas under eaves and sills retaining their colour (Fig. 3.1) but the patchiness can be reduced by the use of a water-repellent treatment containing a toxicant that is effective against staining fungi, although it will be appreciated that this treatment must be particularly resistant to weathering if it is to achieve a reasonable life. In fact there are few components that can achieve the desired resistance to weathering and it is more normal to use a formulation which also contains pigments and binders which give the desired persistent colouration but also leave the natural grain of the wood apparent. Whilst pigment and coating build on the surface of the wood is extremely low compared with conventional coating systems, a high degree of permanence can be achieved through the deep penetration.

FIGURE 3.1 Stain on western red cedar cladding is confined to areas exposed to rainfall and is thus less severe under a window sill. (Penarth Research International Limited)

Although these systems were originally designed as a means for introducing an artificial pigment system in order to maintain the natural colour of wood, a complete range of alternative colours is now available and the systems have become generally known as architectural finishes. The life of these decorative treatments depends largely on their resistance to preferential wetting whilst their colour retention depends not only on the pigments but also on the presence of toxicants which will resist stain development.

In the physical sense, water absorption into wood occurs by capillarity. There is a tension within the water surface and if the water wets the walls of a pore this tension will draw the water into the pore as shown in Figs 3.2 and 3.3. When water wets the surface in this way it is said to have a very small contact angle α. However if this contact angle exceeds 90° the force tending to move the water along the pore becomes reversed and as the contact angle increases further the surface tension tends to repel water from a pore. A water-repellent treatment therefore consists of a coating on the walls of the pores which results in a very large contact angle so that the porous surface can resist water penetration provided that the pressure is not excessive. As the pores remain

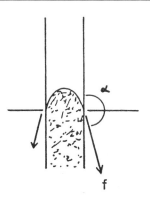

FIGURE 3.3 Water repellency; compared with Fig 3.2 the only change is the contact angle α between the fluid and the solid surface.

open with a water-repellent treatment of this type, trapped water, perhaps introduced by condensation, can disperse by evaporation and it is still possible for the wood beneath the treatment to be affected by changes in the atmospheric relative humidity but only by diffusion along the treated pores. Thus a deeper treatment will result in slower diffusion so that double vacuum or pressure impregnation will invariably give better resistance to seasonal fluctuations than superficial brush, spray or immersion treatments. Pore blocking will always achieve better control over diffusion from the atmosphere and better resistance to seasonal changes in atmospheric relative humidity but dispersion of trapped water will also be restricted.

Wood stabilization

A water-repellent treatment may be effective in reducing absorption of liquid water and deep penetration and pore blocking both reduce the influence of fluctuations in atmospheric relative humidity. If stabilization of the wood is required, both pore blocking and deep impregnation are essential. Resin impregnation alone can be effective as it isolates individual wood fibres from atmospheric changes and physically restrains movement. The effectiveness of resin-impregnation treatment is readily apparent in wooden

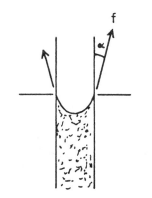

FIGURE 3.2 Capillary absorption resulting from the surface tension within the fluid and the contact angle α between the fluid and the solid surface.

cutlery handles which are now usually impregnated to enable then to resist wetting during washing. There is also increasing interest in the use of resin-impregnated floor blocks; one process, developed in Finland as a peaceful use for atomic energy, involves the impregnation of the wood with monomers which are then cured by exposure to radiation.

Waxes have also been used as impregnants but, while they are quite effective, they are not widely used on their own as they tend to severely affect coating and glue adhesion, although they can be used advantageously at low loadings in conjunction with resins. Polyethylene glycol waxes are also used to stabilize wood but in a rather different way. They are water soluble and generally applied by protracted diffusion into wet wood. When diffusion is complete the treated wood is dried but shrinkage is prevented by baulking by the resin, so that the wood remains in its swollen dimensions. Unfortunately these treatments tend to be hygroscopic, giving a tacky surface unsuitable for the application of decorative paint or varnish coatings. In fact an humectant or hygroscopic agent alone can be used to achieve stability by ensuring that wood is unable to dry and maintaining it in its swollen state. Whilst this type of treatment was developed many years ago it has not been widely used because of the decoration and handling difficulties, but such processes are now being seriously considered in conjunction with chemically reactive surface coatings to overcome these disadvantages.

The wetting of wood as well as the movement with changes in moisture content below the fibre saturation point can be largely attributed to the presence of hydroxyl groups on the cellulose chains, as explained in Chapter 2. Stabilization can be achieved by replacing the hydroxyl groups with other terminal groups or by cross-linking the hydroxyl groups on adjacent cellulose chains. Various systems have been proposed such as acetylation or the use of ethylene oxide,

i-cyanates and organometal compounds but all these systems suffer from the disadvantage that they are completely effective only if the wood is entirely impregnated. It will be explained later in this chapter that complete impregnation is possible only in a limited number of very permeable woods such as birch and it is doubtful at the present time whether many of these treatments are of greater value than the selection of wood of naturally low movement. However if woods of low movement are to be used it is essential to appreciate that they will suffer from non-reversible shrinkage during initial seasoning and they must therefore be thoroughly dried before they are machined or put into service.

Fire

One disadvantage of wood is flammability. Fire will occur only when combustible material is subjected to sufficient heat in the presence of oxygen. In the absence of any one of these three components ignition cannot occur. Within a building it is necessary to withstand fire penetration so that an accidental fire remains isolated for a sufficient period to permit the occupants to escape and to give reasonable time for the fire service to arrive and prevent further damage. Fire resistance is most important where a building is divided into relatively small compartments, but in large open spaces such as corridors, stair wells and roof spaces it is more important to prevent the rapid spread of flame across surfaces. Fire resistance is best achieved in a building by using an adequate thickness of wood in construction and avoiding any minor imperfection which will permit the fire to penetrate through a partition or fire barrier. Clearly, doors should be tight fitting but one solution is to insert a special strip in the edges of the doors or the door frames which is composed of an intumescent material which swells when subjected to fire and thus seals the gaps. The fire resistance of a structure is also improved if the wood is not combustible, and this applies equally to preventing the spread of flame in large spaces in buildings.

Fire retardants

While the provision of satisfactory fire resistance is principally a matter of design, charring rate varies with wood properties, such as density, and wood selection is therefore a significant factor, although charring rate cannot be greatly influenced by treatments. In contrast, fire-retardant treatments can be highly effective in reducing surface spread of flame. An efficient fire retardant acts in several different ways. When the combustible surface is exposed to fire the treatment should prevent flaming and should preferably reduce the rate of degrade and charring. It should reduce ignitability and prevent after-glow, the sustained combustion when heat is removed. If it achieves these requirements it must also be inexpensive, readily applied and permanent with no adverse properties; it must not corrode metal fittings, interact with finish systems or possess unacceptable toxicity.

There are two principal fire-retardant systems in extensive use at the present time. Intumescent coatings act on exposure to high temperatures by foaming into an insulating layer and reducing the rate of temperature rise in the protected wood. This foaming action, which must be completed below wood charring temperature, generally follows a sequence consisting of film softening, bubble formation and ultimate setting into a rigid foam with good adhesion to the surface. Ideally the bubbles should contain inert or preferably fire-retardant gases. Intumescent coatings are really paints and can achieve a decorative function. They can also be applied as a remedial treatment to an existing structure. The use of intumescent strips to seal doors in fire barriers has already been described; these are similar in action to intumescent coatings but are designed to foam to a greater extent in order to achieve complete sealing of

relatively large gaps. Non-intumescent fire-retardant coatings are also available. In principle, they consist of heavy insulating coatings containing components designed to inhibit flaming, perhaps by the generation of water vapour and other fire-retardant gases when exposed to heat.

Impregnation treatments are widely used as alternatives to these coating systems. In principle, these impregnation treatments are similar in effect to coatings and designed to inhibit flaming as well as insulating the wood to prevent deep charring. While their ability to inhibit flaming is as good as or better than that of coating systems, the insulating action is achieved by sacrificing the surface layers of wood and allowing them to char to a limited extent to provide an insulating layer. Unlike the surface coatings which have no further effect once the surface has been physically destroyed, deep impregnation treatments continue to control the rate of charring and prevent flaming and after-glow, even after prolonged exposure to severe fire conditions. A surface coating can be applied virtually universally and will achieve the desired result provided that a sufficient build is obtained, but impregnation fire retardants can function reliably only if adequate retention and distribution is obtained. The reliability of such treatments depends as with wood preservatives on the use of a wood in which adequate penetration can be achieved. Despite these disadvantages impregnation fire-retardant treatments are widely used because structural wood can be pretreated at very low cost compared with that of applying fire-retardant coatings.

Impregnation fire retardants are generally applied by the conventional full-cell vacuum/pressure systems that are described later in this chapter. Generally, these fire-retardant systems consist of aqueous solutions of inorganic salts such as ammonium phosphates, ammonium borates and zinc chloride. All these salts are water soluble and treatments are therefore leachable if exposed to the weather. Where fire retardancy is required in external situations it would appear that resistance to leaching can be obtained by coating with paint or varnish but most fire-retardant treatments of this type are hygroscopic and encourage rapid failure of coating systems through preferential wetting. Polymer fire retardants have been developed that give non-hygroscopic systems which are resistant to leaching and which can reduce natural movement in wood by up to 40%. Whilst formulations of this type are clearly ideal where fire-retardant treatment is required for wood that will be exposed to the weather, they are unfortunately rather expensive and relatively difficult to apply; a two-component system is usually involved which is reacted within the wood by heat application after impregnation.

3.2 Application techniques

Having decided on the active components that are to be employed it is comparatively simple to prepare a preservative formulation but much more difficult to achieve the desired retention and penetration within the treated wood. The most obvious limiting factors are often ignored. There must be sufficient space within the wood to accommodate the desired volume of preservative formulation. This means that the wood must have adequate porosity but it must also have a low moisture content before treatment as this porous space is otherwise occupied by water.

Fluid penetration

A liquid is absorbed into a porous solid by capillarity, a function of the surface tension of the liquid and the angle of contact between the liquid and solid surface as explained earlier in this chapter and illustrated in Fig. 3.2. The capillarity force F which causes liquid flow into pores is given by the formula:

$$F = f \pi d \cos \alpha$$

where f is the surface tension, d is the diameter of the pore and α is the angle of contact. The pressure P developed by this force depends on the cross-section of the pore and is given by the formula:

$$P = \frac{F}{\pi r^2} = \frac{4F \cos \alpha}{d}$$

where r is the radius of the pore. Rate of flow R is given by Poiseuille's formula:

$$R = \frac{P \pi r^4}{8v} = \frac{\pi d^3 f \cos \alpha}{32v}$$

where v is the coefficient of viscosity of the fluid. It is apparent from this formula that the rate of penetration depends particularly on the pore diameter which is a feature of the wood species but it is also clear that a preservation formulation can achieve the most rapid penetration if the viscosity and contact angle are kept to a minimum, with the surface tension as high as possible. It is also apparent form this formula that the penetration rate depends on the ratio of the surface tension f to the viscosity v. As the temperature increases the surface tension decreases but the viscosity decreases to an even greater extent so that this ratio increases with temperature. For example the rate of penetration of water increases by about 25% over a temperature increase of 10°C (18°F). One other interpretation of this formula is that, if the pore diameter is halved, the treatment period must be increased eight times to achieve the same depth of penetration.

Treatment time

Protracted treatment may be unacceptable for economic or technical reasons. For example, if a preservative has a fixation reaction it is probable that, during a protracted treatment period, the active components will be fixed near the surface of the wood and only the carrier sol-

vent will continue to penetrate. Even where a fixation reaction is absent there may still be a tendency for the active components to be deposited at the surface through the preservative solution, migrating outwards during solvent drying. The uniform distribution of preservative components within treated wood is difficult to achieve and the advantageous performance of a proprietary preservative, based on conventio toxicants, is frequently related to the s, solvent and fixation systems that are invo efer t/s, One method that has been suggested for spray= proving penetration is the use of surfactant, large wetting agents, in order to reduce the angl nor- contact, but their use often results in a s arily stantial reduction in the surface tension wl ively actually reduces the rate of penetration. If use of additions are contemplated it is essential spray, they should be evaluated by simple experim; and on wooden blocks. Finally, the coefficien uip- viscosity in this formulation relates to the t vent preservative solution but in organic solv systems, particularly those incorporating res the molecular size of some components may large relative to the pore size so that they unable to penetrate and tend to be filtered screened out of the solution at the surface the wood. If these particular components included in order to achieve a measure of p blocking this screening may be an advant but pore blocking will also obstruct the fur penetration of the preservative solution.

Treatment temperature

The viscosity of some organic preservatives s as creosote increases very rapidly as the t perature is reduced and heating is thus essen if a reasonable penetration rate is to be achiev However, heating the preservative may only limited value as the preservative surf contact with the wood will tend to be by the wood itself and, as it is this par surface that is responsible for the penetr the temperature of the wood will tend to more importance than the temperature

preservative. The wood will be heated by immersion in very hot preservative but this heating process is very slow due to the low thermal conductivity of wood. In cold climates penetration may be achieved more rapidly by slowly heating wood in storage prior to treatment, rather than by using preservative at a very high temperature which may not achieve the same ιισιιу converted green wood but they are much less reliable than simple immersion treatments.

Thixotropic systems

The loadings of flood spray or brief dip treatments depend particularly on the nature of the wood surface but also on the viscosity of the preservative. Higher viscosities result in increased loadings, but unfortunately, improved penetration may not be achieved as it tends to be inhibited by this viscosity increase. One way to improve spray and dip preservative performance is to formulate using a thickening agent which will enable high loadings of preservative

ay tunnel or deluging. This equipment has been largely replaced
ıl Limited)

69

to cling to the surface of the wood but will not obstruct the release of low viscosity preservative solution into the wood, the thickening agent itself remaining on the surface. One example is the bodied mayonnaise-type formulation which has been developed particularly to achieve penetration in remedial treatment. This formulation consists typically of a water-in-oil emulsion in which the emulsion and water system are not essential but are simply a means for obtaining the viscous structure necessary to enable high loadings of preservative to cling to the surface of the wood. The emulsion breaks at the point of contact with the wood, allowing the low viscosity organic solvent preservative to come into contact with the wood and penetrate. Generally, an organic solvent of low volatility is employed in order to avoid evaporation losses during the considerable time that can elapse before the preservative deposit is totally absorbed. There is therefore no restriction on the time required to permit penetration, although manufacturers of preservatives of this type often appear to be unaware of the basic mechanics of the system – if the preservative penetrates very extensively the toxicants must be present at very high concentrations in order to achieve retentions that will be effective.

Immersion treatments

Although these bodied systems are realistic for use in remedial treatment the same principles cannot be readily applied to pretreatment preservatives. The essential feature of the bodied systems is that heavy loadings of preservative are able to remain on the surface until adequate penetration has been achieved, but this necessarily means that there will be difficulties in handling the treated wood during the protracted penetration period. In commercial immersion treatments, it is usually considered more realistic to use simple low-viscosity preservative solutions which will penetrate as quickly as possible, extending the immersion period to achieve the desired penetration. Immersion for

a period up to about 10 minutes is usually described as dipping, whereas longer immersion treatments are known as steeping. Sometimes one hour or more is needed to achieve the necessary penetration, although a low-viscosity preservative may achieve in a few minutes or hours the same depth of penetration as a high viscosity preservative applied for several days. The advantages of immersion treatments are numerous; they are simple, involve low capital coast, and achieve excellent preservative penetration and loading provided that adequate time is available. Immersion treatments are unsuitable for use with rapid-fixing preservatives such as the copper–chromium–arsenic water-borne preservatives and are most suitable for applying low viscosity organic-solvent systems to dry wood.

Diffusion

Diffusion treatment relies on an immersion or spray treatment to load the surface of the wood with a preservative that will subsequently diffuse slowly, to achieve the desired distribution. The best-known diffusion treatment, Timborising, involves the use of a highly soluble borate which is usually applied as a hot solution in order to achieve the required concentration. Freshly sawn green wood with a moisture content in excess of 50% is immersed briefly in the preservative and then close-stacked and wrapped to prevent loss of water by evaporation. A storage period of several weeks or even months for pieces of thick section is required to distribute the borate throughout the wood. This treatment achieves better penetration in very impermeable European whitewood or spruce than any other preservative treatment, apparently because the radial penetration pathways are still open in green wood, whereas they are closed in the dry wood that is typically used for pressure impregnation treatments. Fluoride preservatives are also sometimes applied by diffusion, and some formulations containing bifluorides are said to be suitable for diffusion treatment into dry wood. In fact only the hydrogen fluoride gas diffuses deeply in dry conditions but, whilst it can be readily detected by a reagent when a piece of treated wood is freshly cut, it is progressively lost by volatilization and also by leaching if the wood is exposed to wet conditions.

Osmose process

In the Osmose diffusion process, originally developed for the non-pressure treatment of transmission poles, freshly cut logs are peeled, brushed with a preservative paste and then covered with waterproof paper before being stacked for about three months to allow the salt to diffuse. A variety of preservatives is now used with the Osmose method but the original Osmolit paste was a mixture of sodium fluoride, dinitrophenol and chromates, the latter being included to improve fixation. If there is an insect borer danger, arsenic is frequently incorporated in preservatives that are applied by this method. Very similar diffusion principles are involved in the bandages that can be applied at the ground line, to transmission poles in service, in order to improve protection in this zone where the decay hazard is most severe (Fig. 3.5). Bandages are often based on fluoride

FIGURE 3.5 Pole bandage. The Wolmanit TS impregnated bandage is protected by a weatherproof cover. (Dr Wolman GmbH)

salt pastes which are designed to diffuse into the pole when wet. Tar-oil formulations are also widely used, although they are able to diffuse only when the pole is dry or pretreated with an organic preservative such as creosote.

Boucherie process

In the Boucherie process (Fig. 3.6), preservative is introduced into the sap and is required to diffuse in order to treat the neighbouring zones which are not directly accessible. A cap is attached to the butt of a log immediately after felling and the preservative is introduced under low pressure from a header tank; in this way the sap is displaced and replaced by the preservative solution. Copper sulphate was employed when the process was originally developed by Boucherie but it does not fix within the wood and has now been entirely replaced by multisalt preservatives. Some of the preservatives that are so widely used in vacuum/pressure impregnation, such as the copper–chromium–arsenic (CCA) formulations, fix too rapidly to be applied by this or other diffusion methods, and the more slowly fixing copper–chromium–boron (CCB) and fluorine–chromium–arsenic–phenol (FCAP) formulations are more suitable. In some modern versions of the Boucherie process the design of the caps has been considerably improved by the incorporation of inflated cuffs which permit much higher pressures to be employed, giving more rapid treatment and better control.

Gewecke or Saug–Kappe process

In the Gewecke or Saug–Kappe process a conical cap is fitted on the top end of the log and a

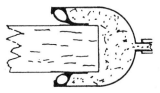

FIGURE 3.6 Improved Boucherie cap with separately inflated cuff to seal the pole butt.

vacuum is applied to remove the sap and induce preservative flow from a reservoir at the butt end. Although this process was developed many years earlier and improved in 1940, it was not until 1950 that Gewecke introduced it into Germany and several years later into Denmark. Normally, the logs are peeled and then fitted with the suction cap at the upper end before being placed in an open tank of preservative. The original Boucherie process would treat a 12 m (40 ft) log in 10–14 days but the Geweche process reduced this time to 1 week and in the improved Gewecke process, with the open treatment tank replaced by a pressure cylinder, the treatment time is reduced to only 20 or 30 hours. In these later processes it is normal to recirculate the sap in the treatment solution, adding preservative at intervals to maintain the necessary concentration, a system that considerably reduces wastage. Originally, the Gewecke process was applied using Basilit UA or preferably Basilit UAS, a more soluble version, although the process is now applied using many other preservatives including the CCA formulation K33.

Hot-and-cold process

In developing countries there is clearly a need for wood preservation in order to improve the economic conditions by ensuring the most efficient utilization of wood and labour resources, but it is equally important to ensure that the process is simple with a low capital cost, even if this results in treatments that are less reliable than those that are considered necessary in more highly developed countries. Dip and sap displacement treatments are currently of particular interest in developing countries but an alternative is the hot-and-cold treatment process. In its simplest form this involves the immersion of wood in cold preservative followed by slow heating which expands the trapped air. When the bubbles of expanding air have ceased the preservative is allowed to cool, thus causing

the remaining trapped air to contract so that the preservative is drawn into the wood. Aqueous preservatives can be applied in this way but if high-boiling organic preservatives are used it is possible to heat the wood beyond the boiling point of water, eventually filling the wood with water vapour. Cooling then results in condensation and the development of an almost complete vacuum which ensures excellent penetration, provided the moisture content of the wood was not too high at the commencement of the treatment process.

This simple form of the hot-and-cold process is widely used by farmers for the butt treatment of fence posts; these are placed in an open drum containing creosote and a fire is lit underneath for heating. If the drum is too full and the fire too hot there is a danger of considerable foaming when the air expands, or even boiling of trapped water, perhaps causing the creosote to overflow and the fire to burn out of control. A more sophisticated version of the process involves the use of separate hot and cold storage tanks for the preservative, with transfer pumps to convey it to an immersion tank, although it is more common to have two separate immersion tanks, transferring the wood between them. In all cases the principle remains the same; air is expanded during immersion of the wood in hot preservative and subsequent immersion in cold preservative causes the air to contract, drawing the preservative into the wood.

Gugel process

In Germany the hot-and-cold method of treatment is known as the Gugel process. In some systems the preservative is replaced in the hot stage by a high-boiling oil such as waste lubricating oil but as there is no appreciable penetration during the hot stage this does not significantly affect the resulting treatment. In Australia the hot stage sometimes consists of prolonged steaming and is followed by transfer to a tank of cold preservative, a process that has been fairly widely used with 3% borax solution to give protection of hardwood against *Lyctus* powder post beetle attack.

Pressure and vacuum treatments

In all treatment methods involving the use of pressure of vacuum it is necessary to place the wood in a pressure vessel, usually known as a cylinder or autoclave (Fig. 3.7). The wood is loaded through a door at one end, usually on railway bogies but sometimes on wheeled baskets, or occasionally loose. In large plants with bogies, it is essential that the wood should be chained down to prevent floating when the cylinder is flooded with preservative as it is unrealistic to pack the cylinder sufficiently tightly for the wood to be restrained. The impregnation process will in any case cause swelling which may cause the load to jam in the cylinder and it is also necessary to leave a small space at the top of the cylinder to permit recovered air to accumulate. Originally the cylinder doors were fitted with bolted flanges but these were very slow to operate and many quick-release door designs are now available which can appreciably improve plant utilization. Loaded bogies are moved into the empty cylinder across a railway bridge which is then removed to permit the door to be closed. One advantage of a bogie system is that the cylinder can be emptied and refilled in a very short period using extra sets of bogies which can be loaded whilst another charge of wood is being treated within the cylinder (Fig. 3.8).

Although it is normal practice to use treatment cycles which will achieve a reasonably dry surface when the charge is removed from the cylinder, some dripping may still occur and it is usual to allow the charge to stand over a drip-collection sump before the bogies are unloaded.

FIGURE 3.7 Typical pressure impregnation plant with bogies for the charge, quick-locking doors, pump house and storage tanks for two different preservatives. (Hickson's Timber Products Limited)

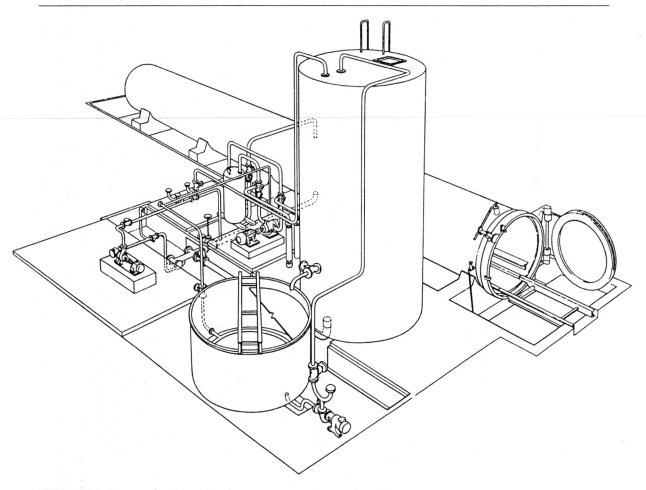

FIGURE 3.8 Layout of a conventional pressure impregnation plant showing the cylinder fitted with quick-locking doors and rails for the feed bogies, the storage tank, preservative mixing tank and pipework. (Hickson's Timber Products Limited)

With water-borne preservatives it is necessary to air- or kiln-dry wood after treatment in order to avoid problems with handling wet wood; whilst the toxicity of these preservatives is important the corrosion of tools and fittings is often a more serious problem.

Rails and bogies waste considerable space at the bottom of a treatment cylinder, perhaps as much as 25% of the volume, and the curved vertical arms which support the load may waste as much again so that the maximum load may be only 50% of the total cylinder volume. This

means that excessive preservative is required to fill these waste spaces during treatment and it is also necessary to use a greater area of steel with a greater thickness to withstand the pressure. In addition, greater energy is needed to move the unnecessarily large volume of preservative and excessive energy is also required to achieve appropriate pressures and vacuums.

One alternative, which is used in a Swedish design known as the 5-T plant (Fig. 3.9), is to use baskets constructed from perforated steel, closely conforming to the internal dimensions of

FIGURE 3.9 A small plant, the 5-T, in which the rail bogies are replaced by baskets to increase cylinder capacity. Baskets are hauled out onto trolleys; a loaded basket can be placed in the cylinder whilst a second basket is being unloaded and reloaded. A trolley system can feed several cylinders. (Anticimexbolagen and Cementone-Beaver Limited)

the cylinder and fitted with small ball-bearing rollers running on the inside of the cylinder. In this way the maximum use is made of the available cylinder space. Unfortunately, the rigidity of the baskets is critical and it is unrealistic to use this design for cylinder diameters in excess of about 65 cm (2 ft 2 in), yet a cylinder of this diameter using baskets gives a capacity equivalent to that of a normal 1 m (3 ft 3 in) diameter cylinder using rail-loading bogies. Baskets are pulled from the cylinder onto a trough for loading, although generally the trough is mounted on a trolley running on rails across

the end of the cylinder, so that several extra baskets are available for loading or unloading whilst a charge is in the cylinder. If it is required to increase the capacity it is more economic to install extra cylinders, served from the same system of rails and troughs, than to install a new large diameter cylinder (Fig. 3.10).

In order to simplify installation the 5-T plants are completely self-contained, consisting of a cylinder mounted on top of a storage tank fitted with the necessary pumps. A treatment installation consisting of two, three or even more of these small cylinders is much less expensive to

FIGURE 3.10 A Gorivac plant with a rectangular treatment vessel. Large rectangular loads can be treated such as packaged wood and completed joinery items, but only vacuum and relatively low pressures can be used. (Gorivaerk A/S)

purchase, install and operate than a large conventional cylinder capable of a similar throughout but there is a limit on the size of the pieces of wood that can be economically accommodated in these relatively small cylinders.

Pressure and vacuum units

Wood impregnation in cylinders can be achieved using a variety of treatment cycles but before discussing these in detail it is necessary to consider the units of pressure and vacuum which are used to describe them. Firstly, it must be remembered that the atmosphere is at a pressure of 1 atmosphere (atm). Drawing a vacuum is an attempt to decrease this pressure to 0 atm. One method to describe both pressure and vacuum is to consider that a complete vacuum has zero absolute pressure so that the atmosphere is at an absolute pressure of 1 atm, and any extra pressure applied on top of atmos-

pheric pressure is additional. Thus the application of 5 atm will result in an absolute pressure of 6 atm, whilst the drawing of a complete vacuum will result in an absolute pressure of 0 atm. This book is intended to be practical and, while it is necessary to interpret some of the more complex treatment cycles in terms of absolute pressure, it is far more convenient to consider the actual plant requirements so that cycles will be quoted in terms of the pressure in atm that must be applied and the efficiency of the vacuum as a percentage that must be drawn. Whilst some perfectionists will object to the use of atm as the pressure unit and percentage as the vacuum unit it must be clearly understood that these are, in fact the only universal units that are widely understood by scientists, technologists and plant operators.

Atmospheric pressure is sometimes described as 1 bar (b), a unit of pressure that gives rise to the familiar millibar (mb) used by meteorologists. Atmospheric pressure is also frequently derived directly from the height of a mercury barometer and described as 760 mm Hg or 30 in Hg. In the metric system pressure is expressed in terms of dynes (dyn) or Newtons (N) per unit area, and for all practical purposes it can be asumed that 1 atm is equivalent to 100 kN/m^2 or 1 000 000 dyn/cm^2. Whilst the current metric standards demand that we should use units involving Newtons, they are still not widely understood and traditional units are still in use at many commercial plants; thus 1 atm becomes, for practical purposes, 1 kg/cm^2 or 15 lb/in^2. The torr has also been fairly widely adopted as a unit of low pressure, particularly vacuum expressed on the absolute scale. A torr is 1 mm Hg, so that complete vacuum is 0 torr whilst atmospheric pressure is 760 torr. In view of the maze of units that are used at present to express pressure and vacuum, the need to confine our descriptions to very simple units, the atmosphere for pressure and the percentage for vacuum, becomes clearly apparent.

Full-cell impregnation

In a full-cell process the aim is to achieve the complete impregnation of the porous spaces within the wood in the hope that a proportion of the preservative will penetrate the surrounding cell walls or that they will at least be protected by the very high loadings of preservative around them. In the empty-cell process the initial impregnation treatment is basically similar but this is followed by a recovery process designed to empty the porous spaces whilst leaving an adequate coating of preservative on the cell walls.

Bethell process

In the traditional full-cell process a sequence of vacuum and pressure is employed to achieve complete impregnation of all the porous spaces within the wood. This impregnation process is currently known as the Bethell method, although it was actually first developed by Breant, and Bethell was responsible only for its adaptation to creosote treatments. In the normal commercial process the wood is introduced into the cylinder and vacuum drawn of 90% or more, the time varying from 15 minutes to several hours depending upon the permeability and cross-section of the wood involved. This vacuum, which removes most of the air from the porous spaces within the wood, is maintained whilst the cylinder is flooded with preservative; water-borne preservatives are generally used at ambient temperatures and warmed only to prevent freezing, crystallization or sludging in cold climates but creosote is usually applied at 60–80°C (140–176°F) to reduce the viscosity and improve penetration. When the cylinder is full the vacuum is released and the preservative starts to flow into the porous spaces in the wood under the influence of atmospheric pressure (Fig. 3.11).

In order to encourage penetration a pressure is then aplied, typically 7–14 atm, and maintained for as long as is necessary to achieve the desired penetration and retention, typically 1–5 hours but occasionally several days, depending on permeability and cross-section. Sometimes

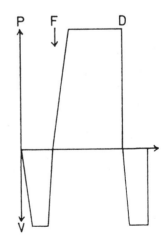

FIGURE 3.11 Bethell full-cell cycle (F = flood; D = drain).

treatment is specified 'to refusal', indicating that the pressure must be be maintained until gauges fitted to the plant indicate that there is no further absorption. With some species of wood, particularly Eucalypts in Australia, much higher pressures are used, but not for softwoods as they may suffer physical damage known as collapse or washboarding if excessive pressures are applied. With some very permeable species of wood the atmospheric pressure on release of the vacuum is sufficient to ensure the necessary penetration or only a relatively low pressure of 1 or 2 atm is necessary; a process involving a vacuum without a superimposed pressure stage is described as a vacuum process whilst one involving a superimposed pressure of less than 5 atm is a low pressure process. After the necessary period the pressure is released and the preservative is removed from the treatment cylinder. Typically, a final vacuum is then drawn in an attempt to remove excess preservative and avoid subsequent bleeding in service.

In theory, this final vacuum is intended to encourage the expansion of any residual trapped air within the wood, forcing excess preservative to the surface where it can drip clear, but in practice the process often leads to excessive surface deposits of high viscosity preservatives such as creosote. A more important function of the final vacuum is perhaps to relieve the compressed state of the wood, thus allowing any excess preservative to be properly absorbed. Whatever the true mechanism, avoidance of bleeding can be achieved with creosote only if heating is maintained throughout the treatment process so that the viscosity of the preservative remains relatively low.

With creosote treatment the nett retention, defined as the loading of preservative that remains after completion of the entire cycle, varies from 80 to 250 kg/m^3 (5–15.6 lb/ft^3) in softwoods, depending on the species, cross-section and the proportion of resistant heartwood and permeable sapwood. In the case of a water-borne salt preservative the nett retention depends on the concentration of the preservative in the solution, but typically 4 to 28 kg/m^3 (0.25 to 1.75 lb/ft^3) is achieved, depending on the preservative involved and the purpose for which the treatment is intended. The Bethell full-cell process is normally used for the application of water-borne preservatives and also for creosote where exceptionally high nett retentions are required in wood for use in extreme hazard situations such as for marine piles. Full-cell impregnation without the use of a superimposed pressure is also normally used in the laboratory for the impregnation of standard test blocks for biocidal evaluation.

Empty-cell impregnation

In empty-cell processes, wood is impregnated with preservative under high pressure on top of air trapped within the wood. This trapped air is later permitted to expand, ejecting preservative from the porous spaces but leaving the cell walls impregnated or coated with preservative. With empty-cell processes it is far easier to achieve treatments that are free from bleeding in service, but empty-cell can be used only when the necessary retentions can be achieved despite the recovery of preservative from the spaces within the wood.

Rüping process

There are two empty-cell processes in common use, both originally designed for use with creosote. The earliest empty-cell process was developed by Wassermann but it is usually named after Rüping who first developed the process commercially. After the cylinder has been loaded and sealed an air pressure is applied, usually 1.7 to 4.0 atm for a period of 10 to 60 minutes depending on the permeability and sizes of the pieces of wood in the charge. The cylinder is then flooded with preservative, usually creosote, without releasing the pressure which is then increased up to perhaps 14 atm, about 10 atm above the original air pressure, and this pressure is maintained until the required gross absorption of preservative is obtained as indicated by the plant gauges. The pressure is then released and the preservative removed from the cylinder, permitting the air trapped within the wood to expand and eject preservative from the porous spaces (Fig. 3.12).

In practice a vacuum of about 60% is drawn during this stage to encourage expansion of the trapped air and to ensure that, despite the relatively high viscosity of the preservative, there are no pockets of trapped air at a pressure in excess of atmospheric. If the pressure is not released in this way there is a danger that the remaining pressurized air will cause continuing bleeding of preservative at the surface of the wood, but the final vacuum will reduce the pressure of trapped air to below atmospheric, so that excess preservative will move inwards when the vacuum is released, giving a particularly clean surface. This final vacuum was not incorporated in the original Rüping process but it is now always used – whatever the empty-cell process it is essential to ensure that any trapped air is under vacuum at the completion of the process in order to avoid subsequent bleeding.

The required gross absorption during the pressure stage is generally defined for individual species of wood, taking account of their permeabilities so that a gross absorption requirement is really a means to ensure adequate penetration. When the pressure is released and the vacuum recovery period completed, a substantial proportion of the preservative will have been removed from the open porous spaces within the wood so that the net retention of preservative may be as low as 40% of the retention from a full-cell process whilst achieving almost as good penetration. For example, in transmission poles penetration is essential, but in most temperate areas a full-cell process is unnecessary with creosote as it will achieve an unnecessarily high retention. Typically, a retention of perhaps 250 kg/m^3 (15.6 lb/ft^3) will be achieved with a full-cell process, but with a Rüping empty-cell process the penetration will be virtually the same but with a retention of only about 110 kg/m^3 (6.87 lb/ft^3). Preservative usage is thus substantially reduced but this nett retention is still adequate to prevent the fungal degradation at the ground line that represents the principal hazard, and the empty-cell process can also achieve freedom from surface bleeding. However, it must be appreciated that good penetration coupled with high recovery and low nett retention can be achieved only with preservatives of relatively low viscosity and this necessarily means that creosote can be used only at high temperatures. In addition, creosote will not satisfactorily coat or penetrate the cell walls if the wood has a moisture content in excess of about 20%.

The Rüping and other empty-cell processes are generally employed for creosote treatments, although they can also be used with water-borne preservatives possessing slow fixation reactions, particularly those that fix only when a component is lost such as the ammonia-based preservatives, which fix as a result of the pH change that occurs when the ammonia volatilizes. Empty-cell processes are also particularly suitable for the application of low viscosity organic-solvent preservatives, achieving excellent

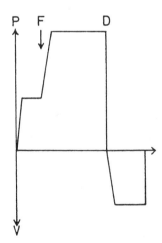

FIGURE 3.12 Rüping empty-cell cycle (F = flood; D = drain).

distribution combined with limited consumption of preservative, although with these low viscosity systems it is unnecessary to use high pressures to achieve the required penetration; a description will be given later of a double vacuum process which is a normal empty-cell process operating with very low pressure differentials.

While Rüping is the most widely used empty-cell process, particularly for the treatment of transmission poles with creosote in Europe, there are a number of other empty-cell processes of importance. The double Rüping process was used on the German railways from about 1909, involving a normal Rüping cycle except that during the impregnation stage a short period of pressure was followed by a vacuum without emptying the cylinder, followed by a return to pressure and the completion of a normal Rüping cycle. The advantages of this modified process are not clear. The additional vacuum would appear to reduce the effect of the initial pressure, perhaps thus improving penetration compared with a normal Rüping cycle but also increasing the nett retention. The process would also seem to have an unnecessarily high energy demand, arising from the application of an initially high air pressure which is later effectively reduced by the application of a vacuum involving the expenditure of further energy. In theory it would seem to be more sensible to reduce the initial pressure alone, but this is effectively the Lowry empty-cell process that was developed in the United States as an alternative to the Rüping process.

Lowry process

In the Lowry process (Fig. 3.13) there is no initial air pressure and the preservative is therefore impregnated on top of air at normal atmospheric pressure. A more intense final vacuum is desirable, perhaps as high as 90%, in order to achieve the maximum recovery but in this respect the Lowry process is never as efficient

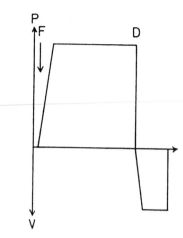

FIGURE 3.13 Lowry empty-cell cycle (F = flood; D = drain).

as the Rüping process; the final nett retention is typically about 60% of the gross absorption compared with as low as 40% with the Rüping process using a low viscosity preservative. Lowry treatment results in less bleeding than the Rüping process because any air trapped at the end of the treatment cycle is at a lower pressure. In addition, a Lowry treatment plant is less elaborate than a Rüping plant as there is no need for a separate air pressure pump. This was at one time considered to be an important economic factor but it is less significant today as many air vacuum pumps can also function as pressure pumps, so that an initial air pressure can be achieved simply at the cost of additional pipe-work and valves.

Nordheim process

The Nordheim process was an adaptation of the Lowry process which attempted to achieve further operating economies. During the impregnation stage the pressure was raised to between 2 and 7 atm and the cylinder valves were then sealed, avoiding the necessity for continuous pumping to maintain the pressure.

In fact the pressure reduced steadily as the preservative penetrated into the wood or through leaks in the plant, giving erratic results so that the process was eventually abandoned.

Energy considerations

It is unfortunate that impregnation processes are often developed by wood technologists with chemical or biological training who usually ignore energy considerations when designing treatment cycles or preparing plant performance specifications. During the impregnation stage it is important to pressurize using a preservative feed pump as this ensures that the level is maintained while the preservative is being absorbed into the wood. A pressure pump of this type need have only low capacity in view of the slow rate of penetration of preservative into wood, and the relative ease with which a fluid can be pressurized in a short period due to its non-compressibility. However, a high-pressure pump of low capacity is quite unsuitable for transferring preservative between the storage tank and pressure cylinder. The pressure pump can be increased in capacity but this also increases the power consumption whilst maintaining the pressure and it may be more economic to provide a second high-capacity pump to achieve rapid fluid transfer.

During the impregnation stage it is essential that the cylinder should be filled with preservative without any air being left at the top as the compression of this trapped air will absorb considerable energy, delaying the pressurization of the cylinder and increasing the cost of operation without achieving any advantage. Pressurizing air or drawing a vacuum in air also requires considerable energy and it is therefore essential to ensure that the cylinder is loaded with the maximum charge that can be accommodated, in order to ensure minimum air space. One possibility is to flood with preservative before drawing a vacuum but, whilst the vacuum can certainly be achieved more quickly, capillary forces between the preservative and the wood prevent the full effect of the vacuum from

being transferred to spaces within the wood; a 90% vacuum above the preservative may represent a 60% vacuum or less within the wood. In addition, simple hydrostatic forces are significant in a large cylinder so that the effective vacuum within the wood is considerably reduced at the bottom of the cylinder where wood is subjected to the hydrostatic pressure arising through the depth below the preservative surface. Clearly, an intense vacuum cannot be achieved within wood whilst a cylinder is flooded with preservative.

Many water-borne preservatives are corrosive but in some commercial and pilot plants direct contact between preservative and pumps is avoided by employing only air pumps. Usually both the treatment cylinder and the storage tank are pressurized so that fluid transfer is achieved by applying vacuum or pressure. A single air pump can be used in this way for all operations but the compressibility of air ensures that the plant can be pressurized only rather slowly and at the expense of considerable energy. Whilst these pumping operations represent perhaps the major operating cost and an area where considerable economies can be achieved by intelligent plant operation, heat energy is equally important in all plants where preservative heating is necessary. Lagging is an obvious precaution but must clearly be extended to all the surfaces of the cylinder, storage tanks and pipework. One interesting concept is to fit the cylinder and storage tanks with water jackets which have a high thermal capacity and thus act as a reservoir of heat energy, enabling off-peak electricity, for example, to be used. The important factor is the temperature of the preservative at the zone where it is penetrating into the wood so that heated preservative can achieve very little effect if the wood is cold. Current practice is to extend the treatment cycle and allow sufficient time for the wood temperature to increase but a more efficient method is to retain wood for a day or two in a heated store before treatment.

High and low pressure

Several variations of these basic treatment cycles are used commercially. The Australian Eucalypts are very impermeable and are often treated at very high pressures to achieve the required penetration; in fact increasing pressure is rather less effective than increasing the treatment time. In contrast, South African pine *Pinus patula* is extremely permeable and only very low pressures are necessary to achieve complete penetration in a short treatment time.

Double vacuum process

In double vacuum treatments an initial vacuum is followed by impregnation under atmospheric pressure with a final vacuum to achieve a degree of recovery and reduce bleeding. The impregnation pressure depends, as in all other processes, on the difference between the pressure applied to the preservative during the impregnation stage and the initial air pressure immediately prior to this stage. In this cycle the initial air pressure depends on the intensity of the initial vacuum but the degree of recovery depends on the extent to which the final vacuum exceeds the initial vacuum, or the degree of expansion of trapped air that can be achieved when the final vacuum is applied. Double vacuum is identical to any other empty-cell process but the impregnation pressure is very low and the process is suitable only for the application of low-viscosity organic-solvent preservatives, in situations where only limited penetration is necessary (Fig. 3.14).

Double vacuum is now extensively used, particularly in Europe, for the treatment of external joinery (millwork) such as window and door frames but it is reliable only if the treated wood is reasonably permeable. It is most often used for the treatment of European redwood or Scots pine *Pinus sylvestris* as the sapwood of this can be readily treated and the heartwood possesses reasonable natural durability. Whilst redwood is most widely used for external joinery in Europe, whitewood, principally spruce *Picea*

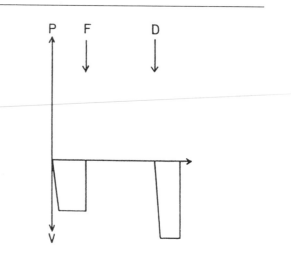

FIGURE 3.14 Double vacuum cycle (F = flood; D = drain).

abies, is preferred in many cases as it is now more readily available through extensive plantings. This whitewood possesses very low permeability and only limited penetration can be achieved by double vacuum treatment, even with preservatives of very low viscosity. Modified treatment cycles with a superimposed impregnation pressure of 1 or 2 atm are often used, a process that is almost identical to the low pressure process that has been used for many years in South Africa, but this higher pressure does not significantly improve the penetration rate and longer impregnation times are much more effective.

Toxicant concentrations

These double vacuum and low pressure processes for external joinery are generally operated under a variety of proprietary names, the first ones being Gorivac in Denmark and Vac-Vac in the Netherlands and the United Kingdom. It is frequently claimed that they provide more reliable treatment of external joinery than the immersion processes which they have largely replaced, but this comment requires some explanation. It is true that short dip will achieve only limited penetration but prolonged immersion can achieve both deep penetration and high preservative retentions. However, the required treatment time may be too long and the high retentions may be uneconomic. The advantage of an empty-cell process is that it can achieve the required penetration at reduced retentions in comparatively short treatment times, thus achieving both reliability and economy. However, it must be clearly appreciated that this does not mean that greater reliability is achieved with a particular preservative if it is applied by a double vacuum empty-cell process, in place of a prolonged immersion treatment which is virtually full-cell, as it is clearly necessary to take account of the nett retentions of preservative within the treated zone in order to calculate the nett retention of the active toxicants that will be achieved. Thus, toxicant concentration in formulated preservatives must be higher when applied by empty-cell processes than when applied by full-cell processes in order to achieve the required toxicant retentions, a factor that is often entirely ignored, and the advantage of an empty-cell process when using a formulated preservative is that it actually achieves economies in the use of carrier solvent without affecting toxicant retentions.

Solvent recovery

Whenever organic-solvent preservatives are used, the cost of the non-functional carrier solvent tends to increase the cost of treatment. There have been many attempts to develop processes which can recover the carrier solvent for re-use. In this sense the various empty-cell processes are largely ignored, although they are certainly the most effective and the most economic means to achieve significant solvent recoveries, and they are the only processes which can achieve recoveries of the relatively non-volatile high-flash petroleum solvents that are so widely used.

For example, it may be required to treat some sawnwood (lumber) with pentachlorophenol at a retention of 6.4 kg/m^3 (0.4 lb/ft^3). A pressure impregnation process on the wood concerned can achieve a gross absorption of about 200 kg/m^3 (12.5 lb/ft^3) and as this is effectively the nett retention when a full-cell treatment is employed it is evident that a 3.2% pentachlorophenol solution is required. The nett retention of pentachlorophenol will be 6.4 kg/m^3 (0.4 lb/ft^3) as required but the nett retention of solvent will be 193.6 kg/m^3 (12.1 lb/ft^3). Alternatively, an empty-cell process can be employed which achieves a recovery of about 50% with a formulation of reasonably low viscosity, so that the nett retention of preservative will be only 100 kg/m^3 (6.25 lb/ft^3) and the formulation concentration must be increased to 6.4% pentachlorophenol in order to achieve the required retention of 6.4 kg/m^3 (0.4 lb/ft^3), but with this process the wood contains only 93.6 kg/m^3 (5.58 lb/ft^3) of solvent.

The use of an empty-cell process can therefore substantially reduce solvent consumption without affecting nett retentions of toxicant, provided the formulation concentrations take account of the degree of recovery that can be achieved, yet the normal Rüping, Lowry and double vacuum processes are seldom considered as means for achieving substantial solvent recoveries. Instead, the emphasis is on alternative processes that utilize relatively volatile solvents which can be employed in their liquid form for a conventional impregnation treatment and later recovered as a gas.

Cellon or Drilon process

In the Cellon process, known in Europe as the Drilon process, the toxicants are applied in a liquified petroleum gas (LPG), normally butane, by conventional full-cell impregnation. If pentachlorophenol is used a concentration of 2 to 4% is necessary, depending on the nett retention required, together with auxillary co-solvents because of the limited solubility of pentachlorophenol in butane. These co-solvents are either low-boiling such as *i*-propyl ether so that they can be recovered with the LPG, or high-boiling such as polyalkylene glycols and retained within the wood. The persistent high-boiling solvents are preferred as they ensure better distribution of the pentachlorophenol within the treated wood and improve resistance to losses through leaching or volatilization.

A vacuum is applied initially to the loaded cylinder and this is followed by purging with an inert gas to remove oxygen. A normal full-cell impregnation cycle follows consisting of a vacuum, flooding with the preservative solution and the application of an impregnation pressure, usually achieved in this particular process by heating the solution with steam coils in order to generate a pressure of 7 to 10 atm. The solution possesses very low viscosity, particularly when heated in this way, and penetration is very rapid. When the required gross absorption has been achieved the preservative solution is removed from the cylinder and a vacuum is drawn in order to volatilize any solvent remaining within the wood. Unfortunately this stage of the process is adiabatic, meaning that the wood is cooled during the initial evaporation stages, reducing the volatility of the remaining solvent. While the generation of pressure, by the temperature increase method, significantly improves the recovery rate it is still necessary to apply a vacuum for a period of 1 to 3 hours before giving a final purge with inert gas and opening the cylinder.

The energy required to maintain the vacuum during the recovery period is expensive and the adiabatic restriction represents a severe limitation which cannot be overcome because of the good thermal insulation properties of the treated wood and the surrounding vacuum conditions. The process seems unlikely to be very economic when these disadvantages are coupled with the time and cost involved in the purging process and the high cost of insuring a pressure plant operating on such low-flash solvents.

Dow process

In the Dow process the butane carrier solvent is replaced by non-flammable methylene chloride which avoids the need for an inert gas purge and considerably reduces insurance costs. In addition, the use of methylene chloride enables the vacuum recovery process to be replaced by recovery in steam so that the adiabatic limitations of the Cellon process are largely avoided. The steam is subsequently condensed, separating the methylene chloride for re-use. However, it is not clear that any significant economic improvement is achieved with this process as the heat energy requirement for steam generation is clearly large. The principal advantage of the Dow process therefore lies in the use of a non-flammable solvent.

Pressure-stroke process

It is clear that these various impregnation processes all have their advantages and disadvantages, and continuous attempts are being made to develop more efficient and more economic processes. Before attempting to develop a new process it is advisable to consider the numerous systems that have been devised in the past and which are now used to only a limited extent or are completely forgotten. The pressure-stroke method is a development of the Bethell full-cell impregnation method in which the cylinder is first flooded with preservative. When the preservative appears in the overflow pipe the top valve of the cylinder is closed but the pump continues to operate, rapidly increasing the pressure. The system is designed with a complex change-over valve so that, after only 3–4 seconds of this pressure stroke, a vacuum can be applied which, within 3 minutes, reaches the vapour pressure of the preservative solution, rapidly expanding air trapped within the wood. After about 15 minutes, loss of air from the wood ceases and a pressure is applied in the usual way (Fig. 3.15).

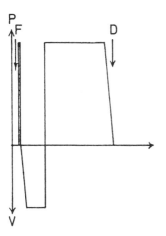

FIGURE 3.15 Pressure-stroke cycle (F = flood; D = drain).

This technique has been largely developed in conjunction with a particular proprietary plant, the 5-T manufactured in Sweden, and the rapid changes between vacuum and pressure are achieved by using a single pump which runs continuously and which generates the vacuum when required by diverting the flow through a venturi device. The method also relies on the fact that the cylinder remains flooded with preservative at all stages; this avoids the heavy energy demands which occur when producing a pressure or vacuum when air is present, although drawing a vacuum in a flooded cylinder reduces the vacuum intensity within the wood as previously explained.

Pressure and penetration

The pressure-stroke method was developed following observations on the rate of flow of preservative into softwoods. If the impregnation pressure is steadily increased the rate of flow increases in proportion until a critical pressure is reached when there is a sudden increase in resistance to flow. It is sometimes suggested that this resistance develops when the increasing flow of preservative through a bordered pit achieves a rate at which the torus is displaced,

effectively closing the pits. It is also suggested that this critical situation develops at far lower pressures in spruce than in pine, thus explaining why only limited penetration can be achieved through pressure impregnation of spruce. Whatever the true explanation it is clear that pressure increases are likely to be of limited benefit and that prolonging the impregnation time is likely to be a more reliable way to achieve penetration into impermeable woods. However, it has also been observed that if a very high pressure is suddenly applied, a very high flow occurs initially, followed by a decrease to the normal expected flow, presumably as the torus becomes displaced.

Oscillating pressure process

The pressure-stroke method is designed to take advantage of this flow characteristic by generating pressures very rapidly and by introducing two pressure stages. The oscillating pressure method (OPM) further develops this idea. The cylinder is flooded with a preservative and then subjected to a pressure of about 7.5 atm alternating with a vacuum of about 95%. Usually the cycles increase progressively in length from about 1 to 7 minutes. The number of cycles varies with the wood, permeable woods usually requiring about 40 cycles and impermeable wood treated in large cross-sections up to 400 cycles. The oscillating pressure method was developed in Sweden and introduced commercially by Boliden in about 1950. In the original version of the process the loading of the cylinder was followed by pre-steaming at a pressure of about one atmosphere, the duration depending on the cross-sectional dimensions of the wood. This pre-steaming was introduced to make the wood more flexible and more amenable to the oscillating pressure method of treatment. A normal oscillating pressure cycle was employed, the total treatment time being 1 hour for every (25 mm) (1 in) of thickness or radius. The process was used originally with S25 salt

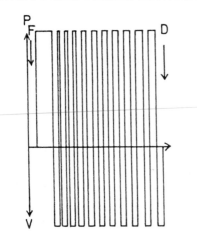

FIGURE 3.16 Oscillating pressure cycle (F = flood; D = drain).

preservative and later with K33. One adaptation of the process is the alternating pressure method (APM), in which the vacuum is omitted and the pressure fluctuates between atmospheric and 10–13 atm (Fig. 3.16).

Poulain process

In transmission poles the most serious decay hazard arises through the action of fungi at the ground line. Any preservation process must take this hazard into account, despite the fact that decay risk is much less on the freely ventilated portion of the pole above the ground. In the Poulain process the poles are first treated to give adequate durability to this exposed portion and then a second treatment is applied to give additional protection to the butt. With this process as originally developed, the poles are first impregnated with a light creosote using the Rüping method then, after the cylinder has been drained, it is rotated into a vertical position and filled with a heavy oil to a depth of about 2 m (6.5 ft) and the air above pressurized to force this oil into the butts. The Poulain method has been used in the Netherlands but with the initial Rüping treatment with creosote replaced by salt treatment, originally by Kyanising but later using Wolman salts – two preservative

systems that are described in the next chapter.

Another system, developed by Kuntz in Hungary before World War II, involved impregnating the poles with a low retention of creosote of lignite tar-oil, then, rotating the cylinder so that the poles were vertical with the butts upwards. The cylinder was next flooded with cold heavy creosote, leaving the butts protruding for about 2 m (6.5 ft). A hot, light creosote oil was then introduced on top of the cold oil and a pressure applied; this caused the light oil to penetrate rapidly into the butts but the heavy oil to penetrate only slowly into the remaining part of the pole, typically giving a penetration of 80 kg/m^3 (5 lb/ft^3) for the butt but only 40 kg/m^3 (2.5 lb/ft^3) for the rest of the pole.

Boulton process

One problem is the limited penetration that can be achieved when wood is wet as, even with water-borne preservatives, there must be sufficient space within the wood to accommodate the required absorption of preservative solution and this means that preservative should never be applied when wood has a moisture content in excess of the fibre saturation point of about 30%. With creosote and other preservatives that are largely immiscible with water, a much lower moisture content is desirable to ensure penetration of the cell wall, although in practice a maximum moisture content of about 25% is usually specified. If the moisture content is higher and kiln-seasoning is unrealistic, as with transmission poles to be treated during the winter months, it is possible to remove water during the treatment process. Creosote is generally heated to reduce its viscosity and, in the Boulton process, this hot creosote is used to boil off the water. Generally, the creosote is heated to about 60°C (140°F) and a vacuum applied to induce boiling. When foaming ceases, pressure is applied as in a normal Bethell full-cell process. When Boulton originally introduced the process, boiling under vacuum was used to avoid the necessity for heating the creosote above 100°C (248°F); some creosotes at that time had very high phenol contents which were appreciably volatile at that temperature, particularly in steam, and it was also feared that high temperatures would damage the wood. Damage does not occur and some treatment plant operators are now using 120°C (248°F), boiling off water during a normal Rüping or Lowry empty-cell process without the need for an additional vacuum stage.

The use of a very hot creosote also helps to reduce bleeding as the viscosity of the preservative is low, achieving both good penetration and recovery, although it must be appreciated that bleeding can be avoided completely only if the cycle is designed to ensure that any trapped air has a pressure below atmospheric at the end of the process so that any movement of preservative will be inwards rather than outwards. In view of the volatile losses that can occur from creosote during Boultonising a modified technique was devised by Rütgerswerke in which the cylinder is first flooded with a separate hot oil for the water-removal stage, if necessary with the application of a vacuum to induce boiling. This heating oil is then removed and creosote impregnated using a normal Rüping cycle.

Preservative bleeding

The normal Bethell, Rüping and Lowry impregnation processes have been used for many years with considerable success, provided that appropriate precautions are taken to ensure that the moisture content of wood is sufficiently low and that the wood to be treated is sufficiently permeable. In recent years the most serious problem has been bleeding, largely because of the complete lack of appreciation that trapped air must have a pressure of less than atmospheric at the completion of an empty-cell treatment cycle.

The most serious bleeding is associated with the Rüping process which involves an initial air pressure followed by impregnation with preservative. In the original process the pressure in the trapped air was relied on to eject excess preservative from the wood but this expulsion would continue, particularly with creosote of relatively high viscosity, for a considerable period after the wood was removed from the treatment cylinder. In many yards, poles were stored for 6 to 12 months to permit the creosote to drip free so that the poles were relatively clean in service. In the Lowry process, in which the preservative is impregnated on top of air at atmospheric pressure, the bleeding has always been less as preservative recovery is achieved through the application of a final vacuum. The addition of this final vacuum to the Rüping process considerably reduces but does not completely prevent bleeding, unless the plant is operating at very high temperatures so that the creosote has a low viscosity and the trapped air pressure can be totally relieved during the vacuum period. In theory, the best technique to avoid bleeding is to use a Lowry process with an extended intense final vacuum or alternatively a modified process involving a limited initial vacuum, perhaps 5 to 10%, with a very intense final vacuum to achieve the required recovery.

These methods to prevent bleeding can be successful only if the preservative possesses low viscosity and with creosote this necessarily means heating. This heating is wasted if the wood is cold or wet as the creosote will be cooled at the critical zone where it is advancing into the wood. Perhaps in the future treatment yards will introduce storage sheds in which wood can be warmed for several days before treatment.

A method to reduce bleeding has recently been introduced in Italy. A normal Rüping process is used to apply creosote but an extended period at atmospheric pressure is introduced, after completion of the pressure impregnation stage and before the preservative is removed from the cylinder. This allows the trapped air to expand and a considerable degree of recovery to be achieved without the necessity for maintaining expensive pumping, the final vacuum after draining more efficiently relieving the pressure of the air trapped within the wood. However, this process can significantly improve recovery and resistance to bleeding only if a substantial time is allowed for the atmospheric pressure stage. Whilst it is true that expensive vacuum pumping is reduced in this way, the extra time involved may considerably decrease plant utilization so that perhaps only two charges can be treated per day instead of three.

Pressure cycle stages

All vacuum and pressure impregnation processes are basically similar and involve five essential stages. The first stage is to load the cylinder and the second is to adjust the pressure of the air trapped within the wood by applying a vacuum or pressure as appropriate. The third stage is to flood the cylinder with preservative and apply the impregnation pressure. The fourth stage involves draining the cylinder and adjusting the pressure of the air within the wood, usually by the application of a vacuum. The fifth and final stage is to return the treated wood to atmospheric pressure then remove it from the cylinder.

Penetration depends on the difference between the initial air pressure and the impregnation pressure, stages two and three respectively, so that it can be improved by reducing the initial air pressure or by increasing the impregnation pressure. The degree of recovery depends on the relationship between the initial and the final air pressures, stages two and four respectively, and the tendency to bleed depends on how these pressures relate to atmospheric pressure. In all cases penetration into relatively impermeable woods can be achieved more readily by increasing impregnation time rather than impregnation pressure; pressures in excess of about 12.5 atm are likely to cause collapse in many woods. In all cases the gross absorption depends on the porosity of the wood and the penetration that is achieved, and the nett retention is

the gross absorption after recovery has occurred in empty-cell processes. In all cases wood must be prepared before treatment. The need to ensure a low moisture content has already been emphasized; with round poles and piles it is also necessary to remove the bark and phloem, and perhaps to peel the pole to ensure a uniform shape. All woodworking should be carried out before treatment.

Impermeable woods

A considerable amount of thought and effort has been devoted to improving the penetration of preservatives into impermeable woods. Although it is clear from theory and practical experience that penetration is not usually improved significantly by the application of high pressures to woods of small pore size, it is a fact that increased pressure remains the favoured method for improving penetration. Pressures in excess of about 12.5 atm are likely to lead to collapse in softwoods but much higher pressures of 50 to 70 atm have been used in Australia for the treatment of relatively impermeable Eucalypt species. Krüzner suggested in 1906 that penetration could be improved by steaming wood for 3 hours at a pressure of 10 atm, equivalent to a temperature of about 180°C (356°F). Whilst an improvement in penetration could be achieved in this way it was far short of his expectations. Two years later Chateau and Merklen adapted the process, relying on high pressure steaming to heat the wood so that a subsequent vacuum stage resulted in the boiling off of trapped water. This process is still used in North America to reduce the moisture content of wood for treatment.

The Taylor Colquitt process was first proposed in 1865 as a means for reducing moisture content by vapour drying but it was Besemfelder who patented the process in 1910. The process relied on passing solvent vapour over the wood – certain water-immiscible solvents such as trichloroethylene and benzene performing best as they were able to absorb water vapour which was later readily removed by condensation. The process was fully developed in 1940 and the first plant became operational in 1945. In the Kresapin process introduced in Austria in 1951, full-cell impregnation with Neckal or Sapro-cresol was followed by drying in the open. The treatments were apparently designed to prevent further water absorption whilst still permitting evaporation to occur; the Kresapin process was followed by normal impregnation with creosote.

Estrade process

The Estrade process, introduced in France and Switzerland for the treatment of spruce involves pre-treatment in the cylinder with hot air. This process induces rapid drying of the external layers of the wood so that substantial splits develop, which encourage deep penetration of preservative. Whilst this system permits the preservative to gain access to the tangential penetration pathways, which are significantly more permeable than the normal radial pathways, the splitting can be a considerable disadvantage through the loss of strength that occurs, particularly in wood suffering from spiral grain.

Incising

Incising can achieve the same penetration advantages whilst actually relieving stresses within the wood and significantly reducing the tendency for splits to develop. Incising is achieved by drilling or forcing needles into the wood, or cutting slits with knives or circular saws; needles are favoured in Germany but knives are generally used in North America and the British Isles. Where slits are cut with knives it is obvious that the best penetration can be achieved by cutting across the grain, to give access to the longitudinal pathways which are most permeable, but this would result in substantial and unacceptable loss of strength so that longitudinal slits, to give access to the tangential pathways, are most realistic. These slits do not need to be continuous as the tangential

pathways give access in turn to the very permeable longitudinal pathways.

Whilst incising was first introduced as a means to improve penetration into round poles, it was adapted in England for sawn spruce using small knives to give a relatively close pattern of slits. When used for this purpose the knives are arranged to penetrate to a depth of about 6 mm (1/4 in) instead of the 15 mm (5/8 in) or more that is normal for the treatment of poles for severe decay hazard conditions. However, it must be appreciated that incising enables treatment to penetrate only to the depth of the incisions. Incising is realistic for Douglas fir, *Pseudotsuga menziesii*, where the treatment of the sapwood alone is required as the heartwood is naturally durable, but the system is less efficient with spruces such as *Picea abies* and *Picea sitchensis* which possess non-durable heartwood. On the other hand, spruce poles often fail through the development of checks penetrating through the treated zone but incising relieves these stresses. Indeed, in woods that are particularly susceptible to checking, a vertical saw kerf is sometimes cut before treatment in order to reduce this danger. One alternative method to reduce checking, in the case of creosote treatments, is to use a cycle that gives a degree of bleeding which will ensure that the surface is coated, and thus sealed, against the wetting and drying conditions that induce that movement within the wood that causes checking.

High energy jets

Prior to the development of incising, even high pressure treatments with water-borne preservatives were unable to effect significant penetration into European whitewood or spruce – a species of increasing importance to the European construction industry in view of the declining availability of European redwood or pine. Only the borate or fluoride diffusion treatments of freshly felled green wood were able to achieve consistent deep penetration but their usefulness is limited as both these treatments are lost in leaching conditions. Needles, drills and knives are not the only means for incising wood as high-energy liquid jets can be used to achieve similar incising patterns. In fact, the preservative can be used as the cutting fluid, achieving incising and preservation in a single operation. This system has been evaluated in the United States but energy costs were found to be very high, apparently because an attempt was made to achieve normal gross absorptions of preservatives. It is much more realistic to achieve low gross absorptions of concentrated preservatives but this is acceptable only with preservative formulations that can subsequently diffuse to achieve reasonable distribution. The depth of the incision depends on the pressure and also on the volume of preservative to be applied to a unit area of wood, so that energy savings can be achieved if the volume can be reduced by the use of a concentrated preservative.

Cobra process

The Cobra process was developed principally as a remedial treatment for transmission poles at the ground line, in order to control incipient decay and extend their service life. The soil is dug away around the pole and a hollow pin is driven in to a depth of about 3 cm (1¹/₄ in). Usually a reservoir is then attached and 2–3 g of preservative solution or paste is forced through the needle. Generally the Wolman type of salt is employed as this is able to diffuse throughout the pole. The process has also been used for the treatment of railway sleepers in service and is sometimes used for the treatment of new wood in special circumstances, such as for treatment with bifluoride to give protection against House Longhorn beetle. The HS-Presser is a similar system for remedial treatment in buildings, particularly in roof structures. A hole of about 8 mm (5/16 in) diameter and 220 mm (8³/₄ in) long is bored into the wood at an angle of about 30°. A hollow needle is then inserted

and fitted with a cylinder containing about 0.7–1.5 litres (1.2–2.6 pints) of preservative solution such as 4% Wolman salt or 10% Osmol WB4. The preservative is usually absorbed in a few days, even in relatively impermeable wood. The Springer-Presser is similar but usually uses 10% Wolmanit or Hydrasil packed in a container pressurized at about 20 atm so that penetration is achieved more rapidly.

Injectors

These injection processes have been simplified by the introduction, originally in France, of small plastic injector nozzles. These are hammered into a previously drilled hole, leaving only a nipple on the surface of the wood. Preservative is then injected at high pressure and the gun removed, a ball valve maintaining the pressure within the wood. If necessary, further preservative can be applied later. This system is ideal for the remedial treatment of external joinery (millwork) such as window and door frames, the preservative spreading for a considerable distance around the injection point but particularly along the grain. When treatment is complete the nipple can be removed and the small hole stopped to conceal the injection point.

Ponding and water spraying

There have been many attempts to develop systems which will enable preservatives to achieve significant penetration into spruce but few have been realistic. One system that deserves particular mention is ponding; if spruce is floated in water for a period of several weeks the permeability of the sapwood is substantially increased, but it must be appreciated that the wood must then be dried before it can be treated. Considerable success has also been achieved, principally in the Irish Republic, with waterspray treatments which are more readily controlled and with the deliberate introduction of bacteria which apparently achieve the increased permeability without affecting the structural strength of the wood. The most serious criticism of this process, at least with regard to spruce, concerns the fact that penetration cannot be achieved in this way into spruce heartwood which is non-durable. On the other hand, in a transmission pole, the strength is only marginally affected by the decay of the heartwood so that this is probably the most realistic method that has been introduced so far for the utilization of spruce in place of pine.

Plywood and particle-board

Wood-based products such as plywood and chipboard must also be treated with preservatives if they are not naturally durable and if they are to be exposed to a deterioration hazard. Plywood can be treated by normal preservation methods in the finished sheet but the glue-lines tend to obstruct penetration. A water-borne treatment increases the moisture content and if the wood veneers possess medium or high movement the individual veneers will swell across the grain, causing enormous stress so that the veneers may separate from the glue-line. The alternative is to treat the veneers before assembly. The simplest method is to spray them with a water-borne preservative as they leave the peeler. Borate preservatives such as Timbor are able to diffuse rapidly into the wet veneer and even rapid fixing copper–chromium–arsenic salts can be applied successfully in this way to veneers of limited thickness.

A further alternative is to incorporate the preservative in the glue-line, a system that is particularly suitable for the application of certain organic insecticides such as Lindane, Dieldrin and Heptachlor (Chlordane) which are appreciably volatile and can become uniformly distributed during the hot pressing stage. These insecticidal glue-line treatments are perhaps the most suitable means for treating plywood so that it conforms with the Australian quarantine regulations but slightly volatile fungicides such as pentachlorophenol can also be applied in this way. Boric acid is not normally considered to be volatile but it becomes volatile in the presence of the steam generated during the hot pressing and it is therefore a very suitable preservative for this process, particularly as it also acts as an accelerator for some adhesives.

The incorporation of preservative components in the adhesive is virtually the only realistic method for treating chipboard. Fibre-board is more difficult to treat as an adhesive is not normally used (see *Wood in Construction* by the present author, page 77). When a fibre-board plant operates as a closed system so that the backwater recirculates, it is possible to dose the backwater with preservative, adjusting the dosage until the required retention is achieved in the final fibre-board product. This procedure is unrealistic with open systems where the backwater is discharged but plants of this type are rare now that there are more severe controls on environmental pollution. Alternatively, a preservative solution can be sprayed either on the wet pulp immediately after straining to form the board or on the completed board, a procedure that is more suitable if the chosen toxicants are appreciably volatile and likely to be lost during high temperature processing. The high-density fibre-boards or hardboards are often 'tempered' by treatment with a drying oil. This tempered board is the only fibre-board product which should be exposed to dampness in service and thus the only product that really requires fungicidal treatment which can be readily achieved by addition to the tempering oil.

Remedial treatments

This description of wood preservation application techniques would be incomplete without a reference to remedial treatment, or the application of preservatives to eradicate established borer attack or fungal decay. Remedial treatment should commence when the damage is observed or suspected. A considerable amount of knowledge and experience of both structures and the wood-destroying insects and fungi is essential to reliably inspect structures and diagnose wood deterioration problems. For example, both termites and House Longhorn beetles can cause very considerable damage before there is any external evidence of their activity, and the Dry rot fungus can spread through plaster, brickwork, masonry and concrete in its search for nourishment and more distant pieces of wood within the structure. The inspection must detect the extent of the damage or, alternatively, suggest the areas which should be opened up to permit a more detailed inspection.

The first task for the treatment operatives is to expose the full extent of the damage in order to decide whether the affected components should be replaced with adequately preserved wood or whether treatment to eradicate the fungal infection or insect attack will be sufficient. This exposure work is perhaps the most important aspect of conscientious remedial treatment as it largely ensures that concealed damage is detected. Preservative treatment then follows. Fungal decay can generally be attributed to a fault in the design, construction or maintenance of the structure which permits wood to become wet, and this fault must be corrected as part of the treatment so that further expertise related to damp-proofing processes may be needed. It is not sufficient to treat only the wood that is visibly affected; adjacent wood may already be infected by a fungus and, in the case of an insect infestation, an attack in one wood component will clearly indicate the hazards faced by all others in the structure.

Remedial treatment must be both eradicant and preservative so that the preservative formulations are often pre-treatment formulations with additional eradicant components. Spray treatment is normally used in conjunction with organic-solvent formulations, which are both more penetrating than water-borne types and also free from the staining that often occurs when, for example, a roof treatment accidentally soaks into a plaster ceiling. Occasionally holes are drilled into beams and other large wood components to permit deep treatment by pressure injection; large beams in ancient buildings are often in contact with damp masonry at either end and therefore suffer from internal decay through the absorption of water by the permeable end-grain. In some cases simple conical nozzles are used fitted to the spray gun but, when pressure on the gun is released, there is a tendency for the preservative to flow out of the injection hole. A more efficient method involves the fitting of an injector system as described earlier in this chapter. Where damage has been caused by wood-borers such as Death Watch beetle it is unnecessary to drill the wood as an injection route has already been provided by the borer galleries. It is unnecessary to inject all flight holes as it will be found that injection into one hole will result in flow from several others. Brickwork and masonry infected by the Dry rot fungus are also drilled to permit proper sterilization, usually with an aqueous formulation as the presence of fungus in the wall indicates dampness which might obstruct the spread of an organic-solvent preservative.

Whilst the stripping of all damaged wood coupled with spray application of preservatives is the most reliable method for remedial treatment, there are less sophisticated methods used in many parts of the world. In Britain, the use of contact insecticide smokes has been recommended as a method for eradication of insect borers such as Death Watch beetle. In fact, smokes do not penetrate into the wood but leave a deposit largely on the upper horizontal surfaces which may kill emerging adult beetles and thus prevent subsequent egg laying. Unfortunately the contact insecticides that are used such as Lindane, Dieldrin and DDT have little persistence when finely dispersed on the surface of wood in this way and these treatments must be repeated at annual intervals for perhaps 8 to 12 years to permit all larvae within the wood to pupate, emerge as adult insects and be killed by the insecticidal deposit. Even prolonged treatment can fail to control an attack of Death Watch beetle in the interior of large beams; this insect is always associated with fungal decay and beams can be hollowed out, as previously described, if their ends are built into damp masonry. Toxic gases, particularly methyl bromide, are also used to eradicate insect infestations but they have distinct disadvantages. The structure must be enclosed within an impervious sheet during treatment and must subsequently be well ventilated to remove the toxic gas. In addition, the gas can kill only insects that are within the wood and the treatment has no preservative action to prevent subsequent reinfestation.

Remedial treatment is extremely complex and a separate science or art. It requires considerable experience in the identification of deterioration hazards, not only those that are of significant economic importance as in wood preservation but also the comparatively rare deterioration problems, described in detail in Appendices B and C, that naturally cause considerable concern when they occur. Knowledge of structures is also essential so that the causes of fungal decay can be clearly appreciated but other remedial treatments may be needed. Remedial wood and related treatments are considered in detail in *Remedial Treatment of Buildings* by the same author.

3.3 Evaluating preservative systems

Service records

The only completely reliable method for evaluating a preservative system is to observe its performance in actual service. Service records are therefore very valuable in confirming the reliability of an established preservation system but alternative techniques are required to realistically evaluate a new preservative system during its development and before its commercial introduction. All acceptable evaluation techniques attempt to reproduce conditions which have been observed in practice to represent severe deterioration risks. There is thus a

danger that these evaluation systems can be too exacting, imposing unnecessarily severe performance requirements on preservatives which may have been designed for use in far less onerous conditions in actual service. For example, a preservative for the carcassing or framing wood of buildings may be required to give protection against fungal decay, resulting from occasional leaks or condensation, or perhaps give protection only against wood-boring insects, and such a preservative does not need to be assessed for performance in severe ground-contact conditions.

Performance classification

Preservatives are classified in this book into four basic groups, shown in detail in Table A.2 in Appendix A, in order to take account of these various hazards. This classification system is not standard but is similar in many respects to the Nordic system, and the required retentions quoted in Appendix A can be readily related to similar classification systems operating throughout the world. Class A refers to wood in normal ground-contact conditions such as transmission poles, fence posts, railway sleepers (ties), piles and structural foundations. It can also be considered to refer to wood immersed in fresh water as in river defence works and even cooling towers, although in the latter case there is an increased risk of Soft rot attack, indicated in the Danish system by a sub-classification Class AS. Such a subdivision is unnecessary as it is now recognized that even ground-contact conditions introduce a risk of Soft rot damage. Class B refers to building and construction wood which is not in ground contact but which is still subject to a moderate risk of decay through accidental leaks or condensation. In many respects the risk is the same as for Class A except that significant leaching conditions are not present. This class includes building carcassing and framing, as well as joinery (millwork) and cladding. Class M refers to preserved wood for marine conditions but applies only when there is a risk of attack by marine borers, particularly gribble *Limnoria* species. Class I applies only when there is a risk of insect attack, particularly by House Longhorn beetle, *Hylotrupes bajulus*, in temperate areas and Dry Wood termites in the tropics. Preservatives meeting this classification generally conform with the Australian quarantine requirements which are designed to prevent the introduction of new wood-borers to Australia. These four classes thus define the most important deterioration risks that preservatives must be capable of withstanding. There are several other service situations which need to be considered such as stain, Pinhole borer and Powder Post beetle control treatments that are used in the forests and mills, but these do not involve standard evaluation techniques.

A Class A preservative in normally assessed throughout the world by the performance of the preservative in actual ground contact in stake trials (Fig. 3.17). Generally, the stakes are comparatively small in cross-section, typically about 50×50 mm (2×2 in), in order to exaggerate natural leaching and the deterioration damage. It is usually considered that the performance of a preservative can be judged reasonably reliably after a period of about five years. Obviously the time factor ensures that this system cannot be used during the development of a preservative.

Laboratory tests

Preservative development normally involves exposure of relatively small blocks of wood to cultures of single fungi in laboratory conditions with decay assessed by the weight loss after a period of perhaps 12 to 16 weeks. This principle is used throughout the world, the tests varying only in the medium on which the fungus is cultured; in the British, Dutch and German systems the fungus is cultured on malt agar

FIGURE 3.17 Simlångsdalen test field in Sweden, one of the sites used to assess Class A preservatives for the Nordic approval scheme.

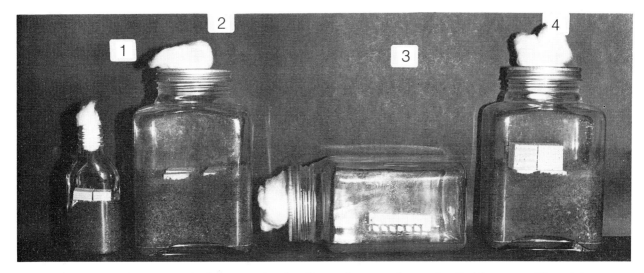

FIGURE 3.18 Laboratory methods for assessing the efficacy of preservatives against Basidiomycetes: (1) a miniature soil and wood block technique suitable for rapid and inexpensive product development tests; (2) the soil and wood block technique used in the American standard test; (3) the malt agar and wood block technique used in the British, German and Dutch standard tests; (4) the soil and wood block technique used in the Nordic standard test. (Penarth Research International Limited)

whereas in the Nordic and American systems the fungus develops on a small untreated block of wood resting on soil (Fig. 3.18). This technique is used for evaluation new toxic chemicals; the test blocks are impregnated with different solution concentrations so that a toxic limit or threshold can be established between the concentrations at which the blocks just decay and those at which the blocks are just free from decay. In this way the preservative activity of various toxicants can be compared and concentrations proposed for their use. When complete preservatives are assessed in this way considerable care is required in diluting them in order to define their safety factor, or the amount of dilution that can be tolerated before decay occurs. Although these tests involve increasing dilutions of the preservative solution and results are therefore obtained as toxicant or formulation concentration, it is normal to take account of absorptions and report the results as retentions in kilogrammes per cubic metre kg/m^3 or pounds per cubic foot lb/ft^3.

Test fungi

The choice of test fungi is also important; for example, *Poria* species are tolerant to copper and should always be used when a preservative formulation contains this element. Generally the test fungi are defined in the appropriate national standards. No attempt is made in this book to describe these standards in detail as they are being continuously revised and it is always advisable to obtain the current standard from the appropriate national authorities.

Weathering resistance

Laboratory block tests can also be used to assess the performance of a preservative after weathering by leaching or volatilization. If a product has good weather resistance and a wide spectrum of activity against the test fungi it can be considered to be a realistic candidate for a full stake trial, although if it is meant to meet only Class B requirements a block test may be considered adequate in many countries, as in the Nordic system and in the British system for the evaluation of preservatives for treating joinery (millwork).

Insect borers

If a preservative is also intended to give resistance to insect borer attack, further tests are necessary against the most appropriate species, these tests usually being carried out in the laboratory. In Europe the House Longhorn beetle represents the greatest danger to structural wood yet protection is also required against the Common Furniture beetle which generally has a greater tolerance to preservatives. Performance against termites is also frequently assessed, the subterranean termite, *Reticulitermes santonensis*, usually being considered the most important species in Europe.

New preservatives

Normally a new preservative is first assessed against a single Basidiomycete fungus such as the Brown rot, *Coniophora puteana*, and if it proves effective at apparently economic retentions the test is then extended to further fungi such as White rot and Soft rot as appropriate to the intended use of the preservative. Leaching and insect borer tests follow so that, ultimately, comprehensive information is available which

clearly establishes whether the new preservative system is likely to be reliable in service. Some tests are unrealistic in the laboratory such as the ground-contact stake test and assessment tests against marine borers; these tests are best carried out in natural conditions where there is known to be a particular hazard. In all these assessments it is normal to compare a new preservative system with a well-known established system, in order to confirm that the tests are realistically severe and in order to provide direct comparisons for commercial reasons. Unfortunately the establishment of preservation reliability represents only a small part of the the time, cost and effort that is required today to establish a new preservative system, health and environmental evaluation being much more difficult. It is therefore not surprising that new preservatives are rarely introduced; only the largest companies and consortia can afford to develop new products today, and most new systems are simply adaptations, based as far as possible on established knowledge and experience. This is unfortunately a situation which encourages the retention of existing products, even if they would not be acceptable if submitted for approval today, a situation that actively discourages the development of more effective and safer products.

Preservation chemicals

4.1 Preservative types

Many different toxicants and other unpleasant substances have been used as wood preservatives. The various preservation systems and their action mechanisms have been described in Chapter 3 but when considering preservative selection for commercial purposes it is first necessary to decide on the basic preservative type that is most likely to achieve the desired function, and it is therefore necessary to establish a reasonably simple and realistic classification for preservatives.

The normal system as adopted in British Standard 1282 involves three main types of preservative. Type TO (tar-oil) comprises distillates of coal-tar including creosote. Type WB (water-borne) includes Wolman salts of the fluoride–chromium type and the copper–chromium formulations which currently dominate this market. The boron diffusion process for green wood is also water-borne but is usually considered to be a special case, as is the use of aqueous solutions of sodium pentachlorophenate in sapstain control and aqueous emulsions of insecticides in Pinhole and Powder Post beetle control. Type OS (organic solvent) involves light petroleum solutions of pentachlorophenol, naphthenates of copper or zinc, chlorinated naphthalenes, organotin compounds and many other less important compounds including contact insecticides. In some areas such as Scan-

dinavia, many of the organic-solvent formulations are decorative and intermediate between a preservative and a paint but often achieving only limited preservative effectiveness.

Tar-oil preservatives are also organic and in some countries such as Denmark it is the practice to include them with organic solvent preservatives when preparing national statistics but there are many other examples of confusing classifications. Thus pentachlorophenol is normally described as a type OS preservative as it is traditionally used in organic solvent carriers but it can be reacted with sodium hydroxide to form sodium pentachlorophenate, which is water-soluble, and concentrated organic solvent solutions can be dispersed in water, as emulsions. The increasing cost of organic solvents and the progressive introduction of more stringent health and safety restrictions have prompted more extensive use of organic preservatives in water carrier systems which should be correctly classified as type WB, but these developments also mean that the distinction between organic solvent and water carrier systems is now less important and no longer an appropriate basis for a preservative classification system. The traditional classification into types TO, WB and OS has therefore been abandoned in this second edition. Tar-oil systems are still considered as an important group but the preservative toxicants or biocides are now classified as inorganic, organic or organometallic compounds, with carrier systems considered in a separate section.

In recent years the health and environmental dangers associated with wood preservation have attracted particular attention. Restrictions on the use of existing preservatives and the requirements for approval of new preservatives have become increasingly stringent, and are now causing serious difficulty to this industry. These changes have not necessarily resulted in improved safety to health and the environment as the development of safer preservative systems is now discouraged by the costs involved and it has been necessary, for economic necessity, to extend the life of established preservative systems which would not be acceptable if they were submitted for safety approval today. These health and environmental safety problems are considered in more detail at the end of Chapter 5.

4.2 Tar-oils

Coal-tar, a product of the distillation of coal, was originally used as a wood preservative but the lighter creosote fraction was later preferred with the introduction of pressure impregnation, as its lower viscosity improved penetration. The development of creosote has been closely associated with the history of the wood preservation industry as previously described in Chapter 1. The word creosote was first used to describe a tar-oil fraction prepared by the destructive distillation of wood and in North America this product is still known as wood-tar creosote. It is now more usual in Europe to describe wood-tar as Stockholm tar and the word creosote is now reserved virtually exclusively for the oil prepared by coal-tar distillation.

The nature of coal-tar varies considerably with the coal type and the processing method. The tar was originally derived principally as a by-product of the manufacture of town gas using horizontal, vertical or inclined retorts with in-termittent or continuous operation at high or low temperatures, about 1350°C (2460°F) or 450–600°C (840–1110°F) respectively. The coal-tar was then distilled to separate the volatile creosote from the residue or pitch, the creosote boiling at 200–400°C (390–750°F) and the pitch boiling above 355°C (670°F); a light creosote has a low residue and conversely a heavy creosote has a high residue.

Creosote availability

It is often said that, as creosote was derived from town gas production, it is no longer readily available. It is certainly true that there is a current world shortage of creosote for wood preservation but this arises through economic factors and is certainly not the result of true scarcity. Coal-tar is still available in enormous quantities from the coke ovens which are associated with the metal smelting industries, but difficulties are encountered in preparing creosote. The old tar distilleries, based in urban areas close to town gas plants, have been closed in many countries as their sources of tar have disappeared and it is not economical to supply them with tar from distant coke ovens. There is no justification for building new distilleries close to the coke ovens while the prices of creosote for wood preservation and other purposes remain low, and it is better to use the tar as a heavy fuel oil in place of petroleum which is becoming increasingly expensive. Poland, the last substantial source of town-gas creosote, has converted to natural gas and coke ovens are now the only source. The price of creosote, depressed for many years while the industry was able to continue to use very old plants with minimal capital investment, is now increasing steadily but this is essential if new coal-tar distilleries or refineries are to be established.

It is interesting to consider the way in which the situation has developed in the United States, as illustrated in Fig. 4.1. Creosote was originally imported into America from Europe as

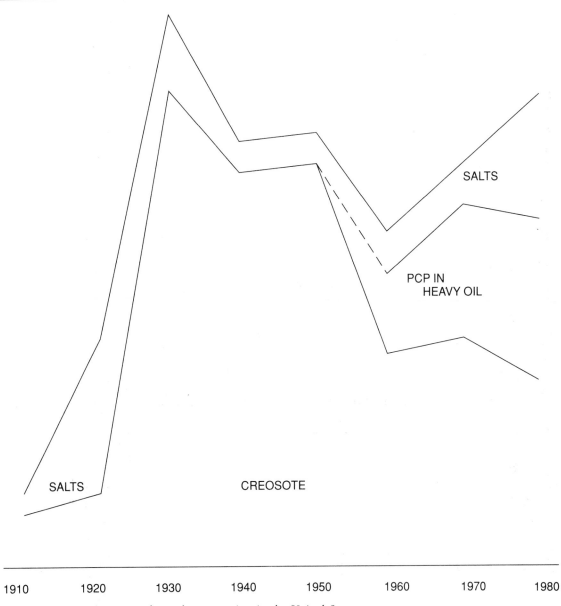

FIGURE 4.1 Development of wood preservation in the United States.

return cargo in petroleum tankers, and considerable shortages were experienced there during World War I, encouraging more extensive use of salt preservatives but also dilution of the available creosote with petroleum. The very rapid expansion of the railways and the trans-mission pole systems carrying power and telephone lines created an enormous increase in the demand for wood preservation after the war and which continued until after 1950, involving particularly creosote treatment. The availability of creosote was decreasing and pentachlorophenol was used increasingly as a substitute for creosote, but ultimately the steel industry established new distillation plants so that the creosote products required by the industry could be produced locally. Creosote and pentachlorophenol have retained a constant volume of the wood preservation market but the market has expanded rapidly since 1960 with the additional volume satisfied entirely by water-borne salt preservatives and by the alternative oxide formulations that are now becoming increasingly popular. The increasing water borne share of the market probably results more from the increasing cost of petroleum rather than any increasing resistance to the use of creosote.

Creosote composition

Clearly, changes in the sources of creosote have resulted in significant changes in composition over the years, particularly in the 19th century with the increasing interest in coal-tar distillation as a source of valuable chemicals. Various specifications have been introduced during the last century in an attempt to standardize creosote and ensure reliability as a wood preservative. It is not the purpose of this book to quote specifications in detail as they are amended progressively and it is far better to consult the appropriate standards organizations when information is required to ensure that only current specifications are used.

The wood-preserving properties of creosote depend on many factors. Coal-tar contains tar acids such as phenol, xylol and cresol, together with tar bases such as pyridine, chinoline and acridine, as well as neutral or dead oil consisting of naphthalene, fluorene, anthracene, phenanthrene, etc. An increase in the high-boiling tar-acid content increases the viscosity whereas an increase in the naphthalene content reduces the viscosity. It was soon recognized with the introduction of impregnation methods that only the use of the low-viscosity creosote fraction can achieve adequate penetration, but it was also appreciated that the fungicidal components

in coal-tar are almost exclusively associated with this creosote fraction.

Creosote as a preservative

The tar acids have excellent fungicidal activity and early creosote specifications therefore emphasized tar acid content. It was presumed that the dead oil possessed no significant preservative properties, and this type of specification continued in use until about 1930. In fact, the wood-preserving properties of creosote depend on many factors and, whilst the tar acids are good fungicides, they are also the components which are most susceptible to loss from wood by volatilization and leaching. A high residue tends to trap tar acids and protect them from loss but it is now appreciated that very high loadings of other less toxic components such as naphthalene are also very important, and high naphthalene creosote has been found to perform best in marine situations. In recent years health risks associated with the use of creosote have been more closely scrutinized, the main risks being identified as the reported carcinogenic properties of polycyclic aromatic hydrocarbons, including the benzopyrenes in creosote. The benzopyrene content can be limited by using only creosote distilling at higher temperatures, a restriction that does not affect impregnation grades but which is now restricting the availability of lighter creosote for surface application for maintenance of fencing. In other respects the use of creosote does not present any unusual health hazards, provided that plants are operated with proper care and persons handling creosote and treated wood observe normal personal hygiene precautions.

The creosote specifications have therefore developed progressively, largely as a result of extensive service experience, but the many conflicting specifications in different countries have introduced unnecessary difficulties. It will be appreciated that the nature of creosote depends on the type of coal used and the nature of the distillation processes so that individual local specifications have tended to be concerned largely with the control of the available product. In 1936 the International Advisory Office of Wood Preservation in The Hague arranged a conference in Copenhagen to agree a specification for use by Scandinavian purchasers. The following year a similar conference was held in Budapest for the benefit of other European consumers, the specifications resulting from these conferences varying only slightly in distillation characteristics and permissible tar acid content so that some creosotes could comply with both specifications. In the United States the American Wood Preservers' Association had developed specifications, and in the United Kingdom the British Standards Institution had specifications for three types of creosote obtained from different sources. Gradually the range of sources has decreased, particularly in recent years as natural gas has replaced coal gas so that all tar is now derived from the coke ovens in the smelting industries. The current specifications ensure reasonable wood-preserving activity combined with reasonable permanence, and are largely based on comparatively recent research on the residues found in wood after long service.

Clearly, long-term performance depends on the heavier and more persistent components but penetration, which depends on the lighter components, is also important. If pressure impregnation is used for creosote treatment it is therefore to increase the temperature, in order to reduce the viscosity of the heavier components and improve their penetration. For example, if the temperature of a typical creosote is increased from 40 to 80°C (100 to 175°F) the viscosity is reduced from 16 to 4 cS. It has been well established in service that after a period of 15 to 20 years in ground contact the failure rate increases. While the failures remain very small in number they are particularly significant in railway sleepers (ties) and transmission poles where the system must be closed if replacements are required. These failures are far less significant

with heavy oil, assuming similar penetration, and the value of heating heavier oil to reduce viscosity is therefore clearly apparent.

Most specifications limit the residue because of the problems that occur through settlement during storage, transport or treatment. Heavy oils tend to bleed more than light oils, although it will be appreciated from the comments in Chapter 3 that this is largely the result of failure to relieve the trapped air pressure in empty-cell processes. In some countries such as the United States creosote continues to be used in combination with coal-tar as it is believed that the tar limits loss of the creosote by volatilization and leaching, and adulteration with petroleum distillates is also permitted in certain circumstances. These mixtures are unusual in other countries where specifications generally require creosote to be used alone, with restrictions on composition to ensure optimum preservative activity and permanence.

Creosote does not perform simply as a toxic preservative as the residue and other heavier components tend to limit moisture content changes so that treated wood is more stable and resistant to splitting. Creosote thus possesses a variety of advantageous properties which are not all readily imitated by alternative preservatives. Formulated organic-solvent preservatives are generally more expensive, particularly if they contain high resin and wax contents designed to achieve this moisture resistance. Water-borne preservatives lack this property completely; attempts to incorporate wax emulsions have not been very successful.

It will be appreciated that creosote possesses many advantages in contrast with the single disadvantage that treatments are relatively dirty. In recent years it has been limited availability that has restricted use rather than any doubts regarding preservative efficacy, yet sufficient coal-tar is certainly available. In 1990 there was sufficient tar feedstock in west Europe for 1 000 000 tonnes of creosote annually compared with a local demand for wood preservation of about 200 000 tonnes and an export demand of about 50 000 tonnes, mainly for North America but also for the Nordic countries and the Middle East. In 1985 half of the European wood preservation creosote was used for railway sleepers (ties) and a third for transmission poles, with most of the remainder used for fence posts. In some countries local variations ensure rather different market divisions; in the United Kingdom there is no significant market for creosote treatment of railway sleepers (ties) as wood has been almost entirely replaced by concrete.

Carbolineum or anthracene oil – Solignum

This section has concentrated so far on creosote for application by pressure impregnation, but a substantial amount is also applied superficially by brush, spray or immersion treatment, particularly on wood that is to be used for fences and other garden or agricultural purposes. Only limited penetration can be achieved and it is therefore essential to use a creosote with good resistance to volatilization and leaching. There are separate specifications in, for example, the United Kingdom, the United States and Germany for creosote of this type which contains a greater proportion of high boiling fractions than the creosote oil used for impregnation. Generally, this creosote is known as carbolineum in continental Europe and anthracene oil in the British Isles and North America. Coloured pigments are sometimes added; the decorative Solignum products introduced into Denmark and the British Isles shortly after World War I were originally based on coal-tar fractions, although modern products contain biocides in petroleum solvents.

Fortified creosote – Carbolineum Avenarius

The preservative activity of anthracene oil or carbolineum is generally similar to that of impregnation creosote and superficial treatments must be particularly generous if effective preservation is to be achieved. Internal decay is not unusual with woods that are resistant to penetration and there have been many attempts to improve reliability. Carbolineum Avenarius was first developed in 1888 and involves the chlorination of carbolineum. In a sense this is the origin of the chlorinated organic compounds that are so widely used in formulated organic-solvent wood preservatives and which are described later in this chapter. Developments in formulated organic-solvent and water-borne preservatives have also suggested several means for fortifying creosote, such as the addition of pentachlorophenol to achieve similar enhancement of the fungicidal properties. Another approach has been to enhance activity by increasing the concentration of components in creosote which seem to be particularly advantageous; one recent development has been the additon of sulphur.

Barol

Barol was developed by Nördlinger in about 1900 and consists of a mixture of copper salts in carbolineum. Combinations of zinc salts and creosote have also been widely employed. Originally the Burnett zinc chloride process was followed by impregnation with creosote; in the United States a petroleum distillate called Bakensfield oil was used. This two-stage process was superseded by a single full-cell impregnation process involving zinc chloride solution suspended in creosote by continuous agitation. Rütgerswerke in Germany developed a mixture of this type with high tar-acid creosote and it was used by the Austrian, Prussian and Danish railways for sleeper (tie) preservation prior to about 1907 when it was largely displaced by creosote alone.

Card process – Tetraset process

The Card process, which was widely used in the United States, involved a mixture of 20% creosote, 2.4% zinc chloride and water maintained in suspension by agitation alone. Despite the low creosote content the preservative properties were almost as good as for creosote alone and the leach resistance was far better than for zinc chloride solution alone. The Tetraset process used in Poland was similar, but a more sophisticated process introduced by the Polish railways shortly before World War II abandoned continuous agitation in favour of a stable emulsion, achieved by using a sulphonate soap and bone glue as a stabilizer; a similar process was used in Croatia during World War II.

These proprietary processes have now been abandoned but two-stage treatments are still used. The Wolman company in Germany replaced the zinc chloride with Triolith salt which is described later, and creosote butt treatment of poles impregnated with water-borne preservatives is still widely used, particularly in the Netherlands with the Poulain rotating cylinder method. Impregnation with a water-borne preservative followed by a non-preservative oil can achieve the reliability of a creosote treatment without its disadvantages; the use of petroleum oil in this way will be described later.

Arsenical creosote

Some creosote such as the vertical retort creosote used in Australia will accept small amounts of arsenic compounds such as 0.46% arsenic trioxide. This addition, introduced in about 1965, has proved very advantageous whenever insecticidal properties are required such as when there is a danger of termite attack. Contact insecticides such as Lindane and Dieldrin have also been used to fortify the insecticidal properties of creosote but they also give additional protection in marine situations against gribble attack which tends to be very resistant to creosote treatment.

Lignite oil – Kreosotnatron

Creosote is normally derived from the fractional distillation of coal-tar produced by the carbonization of bituminous coal, but this is only one of the products of degradation of vegetable matter under anaerobic conditions; the full sequence is peat, lignite or brown coal, lignitous or soft coal, bituminous coal, steam coal and anthracite. Bituminous coal is preferred for the manufacture of creosote as it gives the highest yield of coal-tar but in some areas, particularly in Germany, the most extensive deposits of such vegetable matter are lignite or brown coal. Lignite-oil is creosote derived from lignite-tar which possesses a high paraffin and naphthalene content as well as a substantial degree of unsaturation which results in a tendency for the tar to solidify when exposed to oxygen. Lignite-oil has a very high tar-acid content which was originally extracted with sodium hydroxide to give Kreosotnatron which was used as a wood preservative for pit props in the lignite mines. Tar acids derived in this way were also used to increase the tar acid content of normal creosote when this was considered desirable. The high paraffin content of lignite-oil results in only poor fungicidal activity in the absence of the tar acids and it has usually been used mixed with normal creosote to give good preservation and a clean and dry surface. Most lignite-tar is now used for the manufacture of wax and some resinous compounds.

Wood and peat tar – No-D-K

Peat tar has been used in parts of Russia far from major coal deposits. Peat tar is similar to the tar derived from the destructive distillation of wood. This wood tar can be separated by distillation into various fractions, including wood tar oil or creosote. Softwood tar, often known as Stockholm tar, was at one time extensively produced in Scandinavia and widely used as a brush-applied preservative for log houses and other wood buildings. The preservative actvity is much lower than that of coal-tar or anthracene oil and the treatment was decorative rather than preservative. Wood-tar creosote was also used as a pressure impregnation wood preservative, its low viscosity ensuring excellent penetration compared with coal-tar creosote. Hardwood tar is darker in colour and has been used extensively in wood preservation only in the United States where it has been marketed as No-D-K.

Water gas tar

Water gas is formed by passing steam over hot coke to produce a mixture of hydrogen and carbon monoxide. Water gas can also be carburetted to increase the calorific value by the injection of a petroleum distillate such as gas oil, but this also results in the formation of water gas tar. This process is rare in Europe but more common in the United States. The lower viscosity tar-oil fraction contains similar hydrocarbons to creosote, but without tar acids and with only traces of tar bases and some petroleum hydrocarbons of low activity. Water gas tar oil therefore has limited preservative value but it is often blended with creosote, particularly in the United States where creosote has always been comparatively scarce.

Créosite

Créosite was introduced in Belgium in about 1922. At that time the Belgian railways considered that a high naphthalene creosote was essential for good preservation. In contrast, Créosite was based on the opinion that the naphthalene was virtually non-preservative and that the tar acids, tar bases and neutral hydrocarbons were more important. The naphthalene was therefore removed and replaced by high-boiling hydrocarbons, derived from asphaltic bitumen. The arguments justifying the preparation of Créosite sound like an attempt to justify the removal of naphthalene for some other purpose!

Creofixol – Cornelisol

Creofixol was another special creosote introduced in Belgium in about 1919. This was a normal creosote but with a higher proportion

of low-boiling fractions, giving a low-viscosity penetrating product which was particularly suitable for immersion treatment at ordinary temperatures. Cornelisol was introduced in the Netherlands in 1935 and was a creosote free from tar acids and free from residue, giving a particularly attractive light colour. This preservative has given excellent performance as a treatment for transmission poles.

Transote

Transote was introduced in the United States in 1917 as Reilly transparent penetrating creosote. It consisted of 25–30% refined colourless creosote in a volatile solvent, giving a low-viscosity penetrating product and a treatment that was colourless, free from bleeding and paintable. Laboratory tests indicated that the preservative activity was rather less than might be expected from the creosote content, suggesting that the refining process had removed important components.

Fibrosithe

Fibrosithe was another American product, introduced in 1923 as a creosote formulation for application by brush, spray or immersion. It consisted of 70% creosote with 9.5% coal-tar to give improved surface sealing, 6% phenol to improve the fungicidal properties and finally 8% light coal-tar naphtha and 6.5% benzene to reduce the viscosity and improve penetration. It is perhaps worth mentioning at this stage that low viscosity diluents can significantly reduce the viscosity of a mixture in this way but they do not affect the molecular size of individual components which may still tend to filter out on the surface of the wood rather than penetrate.

Shale oil

Shale oil is derived from the distillation of bituminous shale tar. Bituminous shale is available in many countries including Sweden, Russia, France, Scotland and various parts of the United States. The tar yield from shale is low and processing is justified only when there is a shortage of competitive coal-tar or petroleum products. In Estonia, three types of shale tar have been used as alternatives to anthracene oil or carbolineum and as recently as 1936 shale oil was employed as an alternative to creosote for the impregnation of railway sleepers in Estonia and Lithuania. Tests indicate that shale oil has lower preservative activity than coal-tar creosote but in practice it has performed reliably when used at very high retentions in full-cell impregnation processes, perhaps because the physical barrier properties then become more significant than the toxic properties.

Petroleum distillates

The preservative properties of petroleum oils can be attributed solely to their physical properties as they are virtually non-toxic. In the United States heavy oils are used as diluents for creosote, a practice based on the concept that loadings of creosote can be substantially reduced as in empty-cell treatments whilst still retaining adequate preservative properties, although high loadings achieve greater resistance to weathering. Shortages of creosote in the United States, particularly during World War II, have led to an increase in the use of petroleum-based preservatives, particularly pentachlorophenol in heavy oil, which possess similar properties to creosote. Organic-solvent formulations will be described in detail later in this chapter but it is appropriate to mention this particular formulation at this point as it is directly competitive with creosote. The various toxicants that have been used to fortify creosote have also been applied in heavy petroleum oils in an attempt to develop an alternative to creosote. Copper, zinc and arsenic compounds have been added, but a two-stage treatment involving a water-borne treatment to provide toxicity, followed by an oil treatment to protect against leaching, gives the most reliable results.

4.3 Inorganic compounds

Water-borne preservatives consist of solutions of inorganic ions, originally prepared as mixtures of salts but now often formulated from oxides in order to avoid the unnecessary inactive ions such as sodium and sulphate which are introduced in multisalt formulations. Some inactive groups are important in fixation, so that the loss of ammonia from some formulations results in fixation by a pH change. In salt formulations, an inactive ion may be varied to assist in achieving penetration or the desired retention. Thus sodium fluoride which has limited solubility can achieve an adequate concentration in the normal Wolman-type salts used for pressure impregnation, but it must be replaced by the more soluble and more expensive potassium fluoride to give the higher concentrations that are required when these salts are applied by non-pressure immersion, or when pressure impregnation is used for the treatment of a wood of low permeability.

Multisalt systems – active oxide contents

Originally, simple salts were used as preservatives but they were found to have various disadvantages; many of the most suitable salts were poisonous, corrosive or leachable, or possessed a rather narrow spectrum of activity. All these disadvantages have been overcome by the development of multisalt systems which are now being progressively replaced by oxide mixtures designed to achieve similar combinations of functional ions. There has been an enormous variety of developments and as similar formulations are often known by several different names it is difficult to ensure that any description of water-borne wood preservatives is complete. This account is therefore concerned with the major developments and the most important or most interesting formulations. Where it is necessary to compare multisalt preservatives the current American Wood Preservers' Association system has been adopted.

The most important comparisons are necessary on the copper– chromium–arsenic (CCA) preservatives where the three toxicants are expressed as oxides, the copper as CuO, the chromium as CrO_3 and the arsenic as As_2O_5. The preservative activity of the formulation is then approximately indicated by the total active oxide content, whatever the ratio of the individual toxic elements, although it will be appreciated that the insecticidal properties depend on the arsenic alone and that the ratios also influence the fixation of the various toxic elements.

Mercuric chloride – Kyanising

Mercuric chloride or corrosive sublimate is very soluble and the most important and most effective of the simple salt preservatives. It was first proposed by Homberg in 1705 but the process was patented by Kyan in 1832. Kyanising involved the treatment of wood through simple immersion in a mercuric chloride solution, generally at a concentration of 0.66%. This treatment is very toxic to both insects and fungi but it is also absorbed onto the wood; if wood remains for a protracted period in the solution more mercuric chloride must be added to compensate for this absorption. Mercuric chloride solution is very corrosive and it is necessary to construct the treatment plant from wood, concrete or stone; this is the reason why mercuric chloride has never been applied using an impregnation plant. Mercuric chloride is also very poisonous and very expensive but Kyanising is perhaps the most effective water-borne treatment for European white-wood; spruce transmission poles can be treated by immersion for ten days in mercuric chloride solution and there is still no alternative treatment that can achieve the same degree of reliability! Unfortunately, mercuric chloride was banned as a wood preservative in Germany in 1935 at a time when the Wolman salts dominated the water-borne salt preservative market, but these were insufficiently soluble for use in place of

mercuric chloride in existing Kyanising tanks. A special soluble form of Wolman salt was therefore prepared by replacing sodium fluoride by the more expensive but very soluble potassium fluoride. The progressive development of Wolman salts is described later.

Mixed Kyanising – Deep Kyanising – Chromel – Lignasan

In 1914 Bub developed Mixed Kyanising which involved the use of a mixture of mercuric chloride and copper sulphate with either zinc chloride or sodium fluoride. Deep Kyanising was developed by Kinberg in 1924 and involved steaming and the use of resin solvents such as trichloroethylene to improve penetration. These processes were followed by several attempts to reduce the corrosive action of mercuric chloride. Bryan and Richardson developed a preservative consisting of 1 part mercuric chloride and 2 parts potassium dichromate, a large concentration of sodium nitrite and some sodium carbonate. One product of this type, known as Chromel, gave very good results in stake tests. Although the dichromate was originally incorporated to reduce corrosion so that this product could be used in a normal steel impregnation plant, it was found that it also improved fixation and the product has proved to be very resistant to leaching. Mercury compounds are only rarely used in preservation today. Organomercury compounds were used for many years for sapstain control. Originally ethyl mercury acetate was used but it was soon replaced by the phosphate, a compound marketed as Lignasan, and later by phenyl compounds. The use of these compounds has now been largely discontinued following pollution of streams and lakes through the use of the same compounds for slime control by the paper industry.

Fluorine compounds

Fluorine is historically one of the most important wood preservative elements. Fluorides were originally proposed as wood preservatives in Britain in 1861 in processes designed to precipitate calcium fluoride or fluorosilicate (silicofluoride) in wood but these systems were never developed commercially. The first practical fluoride process, developed in 1901 and patented in 1903, involved dissolving zinc in hydrofluoric acid. The loss of excess hydrogen fluoride after application of the solution to wood resulted in fixation. A similar process developed in France in 1907 involved the treatment of wood with a mixed solution of sodium fluoride and zinc chloride, followed by heating to achieve fixation by the precipitation of zinc fluoride within the wood. The same year Wolman patented the use of sodium and potassium fluorides for the preservation of mine timbers. Sodium fluorosilicate had first been proposed as a wood preservative in 1904 but it had not been widely used, despite its availability and low cost, on account of its low solubility and corrosive properties. In 1926 Wolman developed mixtures of sodium fluoride and sodium fluorosilicate for the treatment of mine timbers, mixtures which were less expensive than sodium fluoride alone, particularly where sodium fluorosilicate was available locally.

Sodium fluoride has been also used as a component in many multisalt preservatives. It is extremely fungicidal and also toxic to some wood-borers such as the House Longhorn beetle, *Hylotrupes bajulus*, although its insecticidal properties are specific to only a few species. Sodium fluoride is non-corrosive and treated wood is paintable, but the treatment is leachable when used alone. The solubility of sodium fluoride is only 4–5% at ambient temperatures but the high toxicity enables adequate retentions to be achieved by pressure impregnation. Potassium fluoride is more expensive but also more soluble and is used when higher solution concentrations are required, such as for immersion treatments or impregnation of woods with low permeability.

Wolman salts – Bellit – Basilit – Malenit

Formulations containing fluorine, chromium, arsenic and phenol components, known in the United States as FCAP preservatives, are usually known as Wolman salts. These formulations originated in Austria in 1907 with a Wolman patent for the use of fluorides for the preservation of mine timbers. In 1909 Malenkovic introduced a preservative consisting of a mixture of about 88% sodium fluoride and 12% dinitrophenolanilin; this preservative was first known as Bellit but renamed Basilit in 1914. In about 1913 Wolman introduced a mixture of sodium fluoride and dinitrophenol which was known as Schwammschutz Rütgers (Rütgers fungicide) as at that time Wolman was associated with Rütgerswerke. Antimony fluoride and zinc fluoride were proposed as additives to reduce the corrosive properties of the dinitrophenolanilin and dinitrophenol, and a mixture of sodium fluoride, zinc fluoride and dinitrophenol was introduced as Malenit in 1921 by Malenkovic.

Flunax – Fluoxyth

A parallel development to these Wolman salts involved a mixture of 84% sodium fluoride with 8% xylenol saponified with 8% of 38% sodium hydroxide solution. This product was known as Flunax or Fluoxyth, and it was usually applied by full-cell impregnation at a concentration of 1.5–3.5%. In 1923 Flunax was used by the German railways for the treatment of pine sleepers (ties), giving a life of 15–16 years compared with 27 years for normal creosote impregnation. Flunax can cause some chemical deterioration, particularly the development of brashness in beech sleepers, so it has not been extensively used, although it was re-introduced during World War II when the shortage of dinitrophenol and chromates limited the production of Wolman salts.

Thriolith – Triolith – Minolith – Thanalith – Tanalith

The true Wolman salts developed following a proposal in about 1913 for the addition of chromates, phosphates or borates to the sodium fluoride and dinitrophenol mixtures as corrosion inhibitors in place of the zinc fluoride which was used in Malenit. Dichromates were found to be most suitable and the resulting product consisting of 85% sodium fluoride, 10% dinitrophenol and 5% sodium or potassium dichromate was very extensively used under the name Triolith, originally Thriolith. This formulation was the starting point for many later improved formulations, the first being Minolith which consisted of Triolith with the addition of a large concentration of rock salt to give fire-retardant properties for wood used in mines. Triolith was designed principally to control fungal decay but in 1922 it was mixed with an equal amount of sodium arsenate to improve insecticidal activity, the resulting product being known as Tanalith, originally Thanalith. Several further patents, principally in the United States, covered similar mixtures but with the dinitrophenol replaced by dinitrocresol or, later, sodium pentachlorophenate.

Dichromate fixation – Triolith U – Tanalith U – Wolmanit U – Wolmanit UA

Although dichromate was originally introduced to reduce the corrosive properties of dinitrophenol it was soon appreciated that it considerably improved fixation and the resistance of the treatment to leaching. In about 1930 the dichromate content in Wolman salts was increased to further improve fixation and these high dichromate versions of the formulations were identified by the suffix U. Preservatives of this type are still used, generally with the following typical formulations:

		Triolith U	*Tanalith U*
Sodium fluoride	NaF	55% (25% F)	25% (11% F)
Potassium dichromate	$K_2Cr_2O_7$	35% (12% CrO_3)	35% (12% CrO_3)
Dinitrophenol		10% (10% DNP)	15% (15% DNP)
Sodium arsenate	$Na_2HAsO_4 \cdot 7H_2O$		25% (9% As_2O_5)

The figures in brackets represent the ratios of active components using the system introduced by the American Wood Preservers' Association, and now widely adopted elsewhere, to simplify comparisons between the ratios of the toxic components in different formulations. Following World War II, Triolith U and Tanalith U were renamed Wolmanit U and Wolmanit UA respectively, for use in Europe.

Trioxan U – Trioxan UA – Wolmanit U hochl. and Wolmanit UA hochl.

In 1934 a proposal was made in Germany to ban the use of very poisonous mercuric chloride. Triolith U and Tanalith U were insufficiently soluble to prepare the relatively high solution concentrations necessary for application by immersion using existing Kyanising tanks, which had been used with mercuric chloride, and special soluble versions of these Wolman salts were developed by replacing the sodium fluoride with more soluble and more expensive potassium fluoride and, at the same time, reducing the dinitrophenol. These new formulations were known as Trioxan U and Trioxan UA, although they were renamed after the World War II as Wolmanit U hochl. and Wolmanit UA hochl. respectively, the suffix indicating that they were highly soluble.

Other Wolman salts

It is useful at this point to consider the current European nomenclature which enables salts of the Wolman type of different manufacture to be identified. The original Triolith became

Triolith U with the addition of further dichromate to improve resistance to leaching and equivalent products are Basilit U, Osmolit U, Wolmanit U, etc. The addition of arsenic to Triolith gave Tanalith, really an alternative name for Triolith A so that Tanalith U is equivalent to Triolith UA, and similar products are therefore Basilit UA, Osmolit UA, Wolmanit UA, etc. Readily-soluble products for use when high solution concentrations are necessary, either for the impregnation treatment of wood of low permeability or for immersion treatments, were achieved by replacing sodium fluoride by potassium fluoride and by reducing the dinitrophenol content to originally give Trioxan U and Trioxan UA, the equivalent products in the Basilit and Osmolit range having the suffix 'leicht löslich', meaning readily soluble, and the equivalent Wolmanit products having the suffix 'hochl', which is an abbreviation for highly soluble. In the most recent development it has been found that the addition of acid gives improved fixation and this has resulted in products known as Wolmanit U (or UA) Reform, now UR (or UAR). In Wolmanit U the fluorine fixation is improved from 20% to 80% whilst in Wolmanit UA the fluorine fixation is improved from 10% to 67% and the arsenic fixation from 75% to 97%. Basilit UAF is similar to Wolmanit UAR.

FCA (or FCAP) salts

Termites represent a risk throughout most of the United States and only the Tanalith or arsenical types of Wolman salts will give reliable protection; they are usually described as fluor–chrome–arsenate–phenol (FCAP) preservatives in America but fluorine–chromium–arsenic (FCA) in Europe. In the United States they are marketed as Osmosalts, Osmosar, Tanalith, Wolman Salts FCAP and Wolman Salts FMP, all conforming with the American Wood Preservers' Association Standard P5 for FCAP preservatives:

AWPA FCAP salts

Fluoride	22% F
Hexavalent chromium	37% CrO_3
Arsenic	25% As_2O_5
Dinitrophenol	16%

Sodium pentachlorophate may be used in place of dinitrophenol

Care is necessary in the interpretation of these American specifications as the percentages are based solely on the active components which are present. In this specification the fluorine, chromium and arsenic are generally present as sodium or potassium salts so that all formulations contain inactive components in addition. A typical commercial formulation will therefore contain much lower concentration of these active components and will be described as only $X\%$ active, depending on the proportion of these components in the product, so that the Tanalith U shown in the earlier table would be only about 47% active; the ratios of the components are also different from those required by the AWPA specification so that the European Tanalith U formulation is not used in the United States.

FCAP pole bandages

Wolman salts of the Triolith or U type are used when fungal decay represents the principal hazard or when an arsenic content is unacceptable. The Tanalith or UA type salts are used where there is significant insect hazard, particularly from House Longhorn beetles or termites. These formulations are generally very effective but severe failures have occurred where they have been used for the treatment of cooling-tower fill, the wooden slats over which hot water runs, exposed to a current of cooling air. This damage was not found to be due not to leaching but to the development of previously unknown resistant fungi which are now known as Soft rots. Some ground-line decay is caused in transmission poles by the same fungi and as a result the Wolman or FCAP salts have been largely replaced for these purposes by the greensalt or CCA preservatives which are described later. Unfortunately greensalts fix extremely rapidly and cannot diffuse to a significant extent before fixation so that, if permeable sapwood is coupled in a wood species with impermeable heartwood, the heartwood can be treated more reliably by Wolman salts as these are able to diffuse into it. This advantage of Wolman salts is particularly noticeable in species of low permeability such as spruce in which heart rot in CCA-treated poles may be more significant than the superficial soft rot in FCAP or Wolman salt-treated poles. If it is required to treat, for example, spruce poles which are both non-durable and relatively impermeable throughout, Wolman salts still provide one of the most reliable treatments available, particularly if the highly soluble salts are used at high concentrations. These highly soluble salts are also used in Osmose, Cobra and other bandage methods for in-situ ground-line treatments to extend the service life of transmission poles.

Berrit – Fluoran OG

Many other fluoride compositions have been developed, usually as imitations of the original salts. These compositions vary in the ratios of the active components but their development was also influenced by materials availability in times of shortage such as during World War II. The serious shortage of chromium compounds throughout Europe resulted in a progressive reduction in their concentration and ultimately their omission or replacement by other corrosion inhibitors. In the Netherlands Triolith was used at first but replaced by Berritt and Fluoran OG, similar to Triolith and Tanalith respectively but with the components in different proportions. Progressive changes continued as individual components became scarce. Eventually, zinc sulphate was used to partly replace the components in short supply, giving so-called

Wolman salts containing about 3 parts sodium fluoride, 3 parts zinc sulphate, 1 part sodium dichromate and a small amount of dinitrophenol. These salts were still in use in the Netherlands in 1950.

Tanalith Un and Tanalith K

The chromium shortage during World War II was the most serious problem and in Rumania Tanalith Un was used which consisted of 30% sodium fluoride, 50% sodium arsenate, 10% dinitrophenol and a reduced chromium content of 10% sodium dichromate. Tanalith K was similar, except that the sodium dichromate was replaced by 5% hexamethylene tetramine and the dinitrophenol by 15% sodium dinitrophenate. This formulation was widely used, although in some areas such as Croatia it often lacked the dinitrophenol content.

Basilit A57 – NAF salt – Fluorising

Basilit A57 was used in Switzerland in World War II, essentially Basilit UA but without the dichromate. A similar product had been introduced into Denmark in about 1935 where it was known as NAF, nitrophenol–arsenic–fluoride. More recently the process of fluorising has been introduced in Australia for the hot-and-cold treatment of karri sleepers (ties) involving a similar formulation of dinitrophenol, arsenic and fluoride. There is virtually no fixation as these formulations lack chromate, but they have been used for the Boucherie treatment of poles where traditional copper sulphate is considered unacceptable because of the danger of development of copper-resistant fungi such as *Poria* species.

Bifluoride treatments

The so-called bifluorides are potassium, sodium and ammonium hydrogen difluoride. Hydrogen fluoride is released when these bifluorides are applied to wood and can diffuse, giving very deep penetration in laboratory tests. Unfortunately this hydrogen fluoride is not fixed and can be readily lost by leaching or volatilization. In remedial treatments this diffusion is particularly valuable and the bifluorides are still widely used in parts of Europe for House Longhorn beetle eradication. The hazards associated with release of toxic hydrogen fluoride do not appear to have been properly considered; reports published in Germany as early as 1940 warned of these dangers, but bifluorides continue to be used. These reports also pointed out that even sodium and potassium fluoride treatments slowly release hydrogen fluoride. Whilst there may therefore be dangers associated with the well-established Wolman FCAP salts which contain fluorides, the bifluorides are certainly an even more serious danger.

Improsol – Rentex

Improsol and Rentex are mixtures of fluorides, bifluorides and chromates, developed as immersion treatments designed to achieve deep penetration. Improsol originally consisted of a mixture of potassium and ammonium bifluoride with wetting agents, and was intended to be applied by immersion at a concentration of 5 to 10% to achieve an average retention of about 1 kg/m^3 (0.06 lb/ft^3) salt. The treated wood was then required to be maintained for about four weeks at a minimum moisture content of 20% to permit diffusion. This treatment had no resistance to loss by leaching and volatilization of hydrogen fluoride. Very ambitious claims were made for Improsol treatment and the formulation has been frequently modified to reduce corrosion and increase peristence.

Mykocid BS – Osmol WB4

Improsol has also been used for sapstain control on freshly converted green wood. Mykocid BS is a similar bifluoride product for sapstain control. These products achieved some penetration of the sapstain market following restrictions on the use of chlorophenols, such as sodium pentachlorophenate, but they are comparatively expensive. Ammonium bifluoride is very effec-tive against stain fungi but it stimulates the development of the mould *Trichoderma viride* which then covers the surface of the wood unless a very high fluoride concentration is employed. The dangers associated with chlorophenols are discussed later but it is difficult to understand the attitude of the Swedish safety authorities in permitting the use of bifluorides whilst restricting the use of chlorophenols which are not known to have caused any mammalian injury in their country. In contrast, regulations in the United Kingdom actually prohibit the use of bifluorides whilst permitting the use of chlorophenols in suitably controlled circumstances, although the bifluoride restrictions were not actually introduced to control wood-preservation treatments. Osmol WB4, introduced in about 1949, is another proprietary sapstain control product based on bifluorides.

Fluorosilicates

Fluorosilicates (silicofluorides) are not as efficient as fluorides as wood preservatives but they are widely available as industrial waste and are therefore a very attractive source of fluorine in many countries. Sodium fluorosilicate was first proposed as a wood preservative as long ago as 1904 but there is still research effort devoted to developing improved systems in Germany, Yugoslavia and Australia. The main problems are relatively low activity associated with low solubility, and corrosion.

Fluorex V and Fluorex S – Sikkuid – Fluralsil – Hydrazil (Hydrarsil) – Basilit CFK

In 1926 Wolman developed a mixture of sodium fluorosilicate with sodium fluoride for the treatment of mining timbers in an attempt to combine reasonable effectiveness with low cost, the fluorosilicate being available from the mineral extraction processes associated with the mining. In the United States sodium fluorosilicate was the principal component in Fluorex V, whereas magnesium fluorosilicate was the principal component in both Fluorex S and in the German product Sikkuid. Another German product Fluralsil consisted of a mixture of sodium fluorosilicate and zinc chloride which was designed to precipitate zinc fluorosilicate. It was originally introduced in 1909 as a wood preservative and for the sterilization of walls infected by the Dry rot fungus, *Serpula lacrymans*, as an early remedial treatment. Fluralsil was also used in Denmark in about 1935, together with the NAF salt that has been described earlier, as an alternative to copper sulphate in the Boucherie treatment of transmission poles, as deterioration by copper-resistant fungi such as *Poria* species was causing concern. Neither Fluralsil nor NAF contain chromates or any other fixation components and both can be readily lost in leaching conditions. Hydrazil, originally Hydrarsil, was a mixture of zinc and mercury fluorosilicates which achieved greater resistance to leaching, although it has now been abandoned because of the toxicity of the mercury content. Copper fluorosilicate is used in Basilit CFK, a product which is described in detail later.

Zinc salts – Burnettising

Zinc has been used for wood preservation in several different forms. Zinc salts were used in combination with fluorides in multi-salt preservatives, particularly as an alternative to dinitrophenol to reduce corrosion and to enhance fungicidal activity and fixation. The use of zinc chloride alone was originally proposed by Burnett in 1838 and for many years Burnettising and Kyanising with mercuric chloride were the most popular salt treatments in Europe. Zinc chloride was used until 1921 in the United States, particularly for railway sleepers (ties), and was still widely used in Russia as recently as 1948. This treatment has poor resistance to leaching but it was found to give excellent performance when used in combination with creosote, originally in a double- but later a

single-stage treatment as described earlier in Section 4.2. The use of zinc chloride followed by creosote economized on the use of creosote but achieved virtually as good protection as higher retentions of creosote alone. This double treatment was used by the Danish railways from 1889 to 1907 when economic factors caused it to be abandoned in favour of creosote alone. The toxicity of the creosote is unnecessary and this treatment can therefore use non-toxic petroleum oils, a useful alternative in times of creosote shortage. In the Card process in the United States and the Tetraset process in Poland the two-stage treatment is replaced by a mixture of salt solution and creosote, maintained in suspension by agitation or emulsifiers; these processes were described in detail in Section 4.2.

Kulba salt – chromated zinc chloride (CZC) – copperized CZC (CCZC)

The excellent solubility of zinc chloride is helpful in the preparation of the solution but also results in poor resistance to leaching. The corrosion of ferrous fittings can usually be attributed to the presence of hydrochloric acid. Although zinc chloride is not so active as some other salts it is still used as a component in several multi-salt preservatives. One of the simplest is Kulba salt which was introduced in Belgium during World War II and which consists of a mixture of zinc chloride and sodium hydroxide. Chromated zinc chloride (CZC) was developed in the United States in about 1934 and achieved considerable resistance to leaching. It consists of zinc chloride and sodium dichromate mixed according to ratios defined in the American Wood Preservers' Association Standard P5 which requires the hexavalent chromium content to be 20%, as CrO_3, and the zinc content to be 80%, as ZnO. In copperized CZC (CCZC) about 10% of the zinc chloride is replaced by cupric chloride, to broaden the spectrum of activity.

ZFD – ZFM

Zinc sulphate has also been used as a component in several multisalt preservatives. ZFD preservative was used in Belgium in World War II and consisted of a mixture of zinc sulphate, sodium fluoride and dinitrophenol, whereas ZFM was zinc sulphate and magnesium fluorosilicate (silicofluoride). In the Netherlands zinc sulphate was added to Triolith when it was in short supply, resulting in a product that was still used in 1950, but when Triolith was completely unavailable zinc sulphate was used alone. Large quantities of zinc sulphate are available from the Witwatersrand mining operations in the Transvaal and this was used from about 1918 in a mixture with Triolith as a preservative for the treatment of mining timbers; it was normally prepared using 3% zinc sulphate and 0.3% Triolith, and replaced zinc chloride which had been used previously. High loadings were necessary to achieve an appreciable preservative effect, partly because zinc sulphate has poor resistance to leaching, and this formulation was suitable only for application to permeable wood.

Zinc meta-arsenite (ZMA) – Boliden BIS – Boliden S – Boliden S25

Zinc meta-arsenite (ZMA) was developed as a wood preservative by Curtin in the United States in 1928. It is applied to wood in an acetic acid solution but, as the acid is lost by volatilization, a non-leachable precipitate of zinc meta-arsenite is formed. This process was extensively used in the United States before World War II. Zinc was also used in a number of the early Boliden salts developed in Sweden. Boliden BIS was a zinc, arsenic and chromium salt mixture. Boliden S was similar in composition but prepared as an oxide mixture. In Boliden S25 about 25% of the zinc was replaced by copper, but all these mixtures were eventually superseded by K33, a copper–chromium–arsenic oxide preservative; the development of Boliden salts will be described later. In Australia a copper–chromium–zinc–arsenic salt mixture is still used which is similar in composition to Boliden S25 but prepared from salts rather than oxides.

Copper salts

Copper is the most important component in most of the relatively modern preservatives. Copper sulphate was originally proposed as a wood preservative by Margary in 1837 and was used extensively, usually as a 1% solution, although it could not be applied in hard water which caused precipitation. Copper sulphate possesses high fungicidal activity but is very corrosive to iron and steel. Copper impregnation equipment was sometimes used but it was very expensive. Copper sulphate is very soluble but, while most of the treatment is leachable, a proportion remains fixed within the wood. The most realistic use of copper sulphate was in transmission pole treatment by the Boucherie method – a cap was fitted to the butt of a freshly felled log and the copper sulphate solution was introduced under low pressure from a header tank, gradually displacing the sap. This sap-displacement process was extensively used in France and later introduced into other countries such as Denmark and Finland; Boucherie and similar processes are still in use with various preservatives. The performance of copper sulphate-treated poles was reported to be poor in alkaline soils but the failures were actually associated with soils containing ammonia rather than with carbonaceous soils in which the copper is converted to the carbonate which is still fungicidal, but resistant to leaching. The main problem with all copper preservatives has been that some fungi are resistant, particularly the important *Poria* species, which are able to detoxify the copper by formation of the oxalate. Copper-based preservatives must therefore contain other toxic fungicides if reliable preservation is to be achieved at reasonable retentions, although it is interesting to note that the arsenic incorporated as an insecticide distinctly improves activity against resistant fungi.

Aczol – Viczsol

Aczol, known in Germany as Viczsol, an ammoniacal solution of copper and zinc salts with phenol, was originally introduced in 1907 and was progressively improved during the following thirty years. After treatment, the loss of ammonia results in good fixation but it has been found in laboratory tests that the effectiveness of the product declines steadily, as the fixation progressed, so that realistic assessments can only be made after a sufficient period to allow for complete fixation. Early test results exaggerated effectiveness and there were initially some failures in service through the use of inadequate retentions. It was also reported that this treatment made wood rather brittle.

Chemonite – ammoniacal copper arsenite (ACA) – ammoniacal copper–zinc arsenite (ACZA)

In Chemonite the copper is used without zinc but arsenic is added to improve the insecticidal properties. Chemonite originated in the United States in about 1925 when Dr Aaron Gordon of the University of California proposed that Paris green, copper aceto-arsenite, should be used as a wood preservative. Chemonite, which was first marketed by the Diamond Match Company in 1934 and from 1935 by the Chemonite Wood Preserving Company, was originally a mixture of copper, arsenic and ammonium acetates, but it is now usually prepared as a 6% solution for wood preservation by mixing 1.84% copper hydroxide and 1.3% arsenic trioxide with ammonia and small amounts of acetic acid and glycerol, although different copper compounds are used at times. This formulation, in which the ratio of copper and arsenic is 49.8% CuO to 50.2% As_2O_5, is known in America as ammoniacal copper arsenite (ACA). Half of the arsenic As_2O_5 content has been replaced since about 1983 by zinc as ZnO, a version known as ammoniacal copper–zinc arsenite (ACZA) which has now replaced the original ACA formulation in all plants in the United States. The copper and arsenic are fixed, mainly as oxides, as the ammonia evaporates; this slow fixation process is very advantageous in the treatment of impermeable species and also enables Chemonite to be applied by the Boucherie and similar sap-displacement processes. There is no corrosion of steel plant but the treatment corrodes copper. Chemonite is generally very reliable, although it has been reported that it may be less effective in marine situations.

CAA – ZAA

More recently, the Canadian Forests Products Laboratory has developed ammoniacal preservatives which are described as copper-ammonia additive (CAA) and zinc–ammonia additive (ZAA), although it would seem more sensible to attribute these initials to copper–ammonia-arsenic, as the formulations contain arsenic as an insecticide. These preservatives can achieve exceptional penetration and are thus considerably more reliable than the well-established copper–chromium–arsenic (CCA) preservatives for the treatment of impermeable species. However, difficulties have been encountered in substantiating these claims and it seems more likely that the main advantage of the CAA and ZAA preservatives, as well as of the older-established ACA (Chemonite) product, lies in their slow fixation and the protracted diffusion that can occur, resulting in deep penetration in suitable circumstances. The use of arsenic in a recently developed product was surprising and Domtar in Canada soon developed similar products with the arsenic replaced by quaternery ammonium compounds; these formulations are known as ammoniacal copper–quaternery (ACQ) or zinc–quaternery (AZQ) formulations.

Ammoniacal copper borate (CAB) – acid copper borate (ACB) or acid zinc borate (AZB)

Increasing resistance to the use of arsenic also prompted the development of ammoniacal copper borate systems in about 1965 but they have not been used commercially to a significant extent, apparently because the various less expensive and well established arsenical systems have continued in use in many countries, despite health fears. Fixation by loss of ammonia is not essential to copper or zinc borate systems and much simpler formulations are possible comprising copper or zinc acetate with boric acid. Different ratios can be used to form various borates, the zinc borates having well-established fungicidal and fire retardent properties; they are often used as active pigments in plastics and paints.

Another ammoniacal formulation involves copper carboxylate but it is essentially a water-solubilized copper soap; this system will be described more fully in Section 4.5.

Copper–chromium–arsenic (CCA)

One of the most important advances in wood preservation was the development of the copper–chromium–arsenic preservatives which are known in the United States as chromated copper arsenate (CCA). There were three independent development routes that ultimately yielded basically similar CCA products.

Acid copper chromate (ACC) – Celcure N – Celcure A – Celcure F

Celcure was developed by Gunn in Scotland in 1926 but was then improved progressively up to 1942. The original Celcure product, now sometimes known as Celcure N, is described in America as acid copper chromate (ACC). It consisted of a mixture of copper sulphate, a dichromate and acetic acid, but the acetic acid was replaced with boric acid and phosphates and zinc chloride were added to a fire-retardant version, Celcure F. The product was virtually free from corrosion and highly fixed, giving excellent protection against all fungi except a few copper-resistant species, particularly *Poria* species, but its performance was unreliable against insects and crustacean marine borers,

even at high loadings. Arsenic was therefore added to form the CCA product Celcure A which gives excellent performance in all respects; this final product will be discussed later in comparison with other CCA formulations.

Arsenic salts

A second route of CCA development, in Scandinavia, involved particularly Boliden products. These preservatives were originally developed as a means of utilizing arsenic waste from the iron-ore industries in Sweden. Arsenic is a poison and this has always caused difficulties when it is used for any purpose; the dangers associated with arsenic preservatives will be described later. Arsenic was first proposed as a wood preservative by Baster in 1730 and there have been several proposals since, principally for preservatives against insect attack. Some fungi are also controlled by arsenic but others are tolerant and can convert arsenic deposits to the toxic gas arsine. There is no significant danger of arsine production in wood preservatives, provided that arsenic is used only as a component in formulations which contain other fungicides capable of controlling these arsenic-tolerant organisms.

Boliden BIS

Arsenites, derived from arsenic trioxide As_2O_3, are less stable and less soluble than arsenates, derived from arsenic pentoxide As_2O_5. Sodium, copper and zinc arsenites have been proposed as wood preservatives, as well as sodium, copper, zinc and chromium arsenates, but all have now been superseded by modern multicomponent preservatives. The first Boliden preservative system was developed before 1932 by Stålhane as a two-stage process; impregnation with sodium arsenite solution was followed by a second impregnation with zinc chloride solution, a double decomposition resulting in the precipitation of zinc arsenite. The sodium arsenite was soon replaced by the more soluble arsenate but the process was rather unreliable. Drying was necessary after the first stage to ensure sufficient free volume within the wood to accommodate the second-stage impregnation. An empty-cell treatment was considered for the first stage followed by a full-cell treatment for the second stage but the cost of double treatment was excessive and the process was abandoned in 1936, to be replaced by a single-stage process developed by Häger. In this process the reducing properties of the wood were used to change the valency state of the dichromate in the formulation, in order to achieve fixation of the zinc and arsenic components – by formation of chromates and by attachment to the wood structure. This product was known as Boliden BIS salt and was still in use in Sweden in 1950. It consisted of a mixture of arsenic acid, sodium arsenate, sodium dichromate and zinc sulphate, the proportions varying slightly over the years of use. Boliden BIS generally performed well in service but there were some failures and the product was considered to be unreliable.

Boliden S – Boliden S25

Häger therefore developed a new version of the formulation in which he replaced the salts by oxides, omitting virtually all the non-active parts of the formulation and achieving a more toxic product which was consequently less expensive to transport. This substitute for the Boliden BIS salt preservative was known as Boliden S and was prepared as a paste by mixing zinc oxide and chromium trioxide into arsenic acid (arsenic pentoxide solution). 25% of the zinc was replaced with copper in Boliden S25 to further improve the preservative activity but, although both these preservatives were given extensive trials in many parts of the world, they were soon superseded by K33 in which all the zinc was replaced by copper.

Lahontuho K33 – Boliden K33

K33 was first marketed in Finland in 1949 as Lahontuho K33 and was introduced in Sweden as Boliden K33 the following year, eventually becoming one of the most widely used CCA preservatives. It was normally manufactured as a paste by mixing copper oxide or carbonate and chromium trioxide with arsenic acid (arsenic pentoxide solution). It was sometimes produced as a concentrated dry powder to reduce transport costs but the paste version was preferred as it avoided toxic dust problems during the preparation of preservative solutions. In some areas close to production plants it was also produced as a bulk solution, normally equivalent to 60% of the normal K33 formulation. K33 was marketed by several different companies throughout the world; it is known as Lahontuho K33 or Häger K33 in Finland, as Boliden K33 in Sweden, and as Boliden CCA and Osmose K-33 in North America. There are also a number of other preservatives which possess similar toxic component ratios such as Koppers CCA-B in North America, the latter title reflecting classification of formulations of this type as CCA type B in American Wood Preservers' Association Standard P5, as explained later. CCA-B formulations have been largely withdrawn in recent years and replaced with lower arsenic systems, mainly formulations conforming to CCA-C and British Standard BS 4072 requirements.

Falkamesan

In about 1930 Falck and Kamesan developed a mixture of arsenates and dichromates. It will be appreciated from earlier comments on Wolman salts that dichromates were originally incorporated in preservatives as corrosion inhibitors but that it had been later observed that they often improved leach resistance. In this particular development it was claimed that the dichromate was included solely to improve fixation. The ratio of arsenate as As_2O_5 to dichromate as CrO_3 was studied and the optimum resistance to leaching was found to occur at a ratio of about 1:0.65 and this ratio was therefore used in Falkamesan preservative. This product could be prepared as a mixture of sodium arsenate and potassium dichromate, or alternatively arsenic pentoxide and chromium trioxide if it was required to increase the active concentration and reduce transport costs. The potassium dichromate could be displaced in theory by less-expensive sodium dichromate but the sodium salt is hygroscopic and the potassium salt was normally used to ensure a free-flowing powder formulation; these comments apply to all preservatives of this type.

Ascu – Greensalts (CCA) – Erdalith

Falkamesan had been developed mainly as a preservative against termite attack but in about 1933 copper sulphate was added to improve activity against fungi, resulting in Ascu, the first true copper–chromium–arsenic wood preservative. In its original form Ascu was prepared by dissolving one part arsenic pentoxide dihydrate with three parts copper sulphate pentahydrate and five parts potassium dichromate in 200 parts water, to give a preservative for application by normal full-cell pressure impregnation. The treatment gives the wood a distinctly greenish colouration and the preservative soon became known as greensalt, a general term that is now applied to all CCA products, although it has also been adopted as a trademark in the United States. Ascu was marketed in the United States as the original Greensalt, or Greensalt K, which incorporated potassium dichromate and which was first used by Bell Telephones in extensive experiments in 1938 in an attempt to develop clean and reliable pole preservation treatments. Greensalt S was identical but the potassium dichromate was replaced by an equivalent amount of sodium dichromate; this product was also known as Erdalith. Greensalt O was an alternative formulation, patented by McMahon in 1948, and prepared by mixing 2.9 parts chromium trioxide with 1.35 parts of alkaline copper carbonate and one part arsenic pentoxide dihydrate, giving approximately the same ratio between copper and arsenic but with a slight increase in chromium.

Celcure A – Tanalith C

Industrial expansion in the United Kingdom after World War II was dependent on a substantial expansion in power generation capacity which involved the rapid construction of new power stations, many equipped with natural-draught water-cooling towers. In these towers water is sprayed over a stack (or fill) of specially shaped and positioned wood slats, to expose the hot water to a rising draught of cooling air, the flow being induced naturally by the parabolic shape of the tower. The fill is thus continually exposed to warm water. In addition, there is often a second layer of slats known as the mist eliminators higher up the tower above the sprays to prevent droplets of water from being carried out of the tower and causing winter icing on neighbouring roads. These mist eliminator slats generally possess a lower moisture content than the fill. There is clearly a severe decay risk and cooling tower wood was originally treated in the United Kingdom with the ACC preservative Celcure or the FCAP preservative Tanalith U. Deterioration of the Tanalith fill slats developed rapidly but it was soon discovered that this was not due to excessive leaching, but to the presence of previously unknown fungi which were named Soft rots as they attacked the external surfaces of the wood, causing softening to a progressively increasing depth. There were also reports of ground-line failure in Tanalith U transmission poles in South Africa, again apparently due to soft rot attack. In contrast the Celcure fill slats were in sound condition, although some of the mist eliminator slats were decayed by copper-resistant *Poria* fungi. The British manufacturers of Celcure and Tanalith U were aware of the reports of the excellent performance of greensalt treatments in the United States and both introduced new treatments of this type. Celcure became Celcure A through the addition of arsenic which improved effectiveness against insects as well as against copper-resistant fungi. Tanalith U was replaced with Tanalith C, an entirely new CCA preservative, developed following extensive studies of the Bell Telephone trials with greensalt preservatives based on the original Ascu formulation.

CCA preservatives

CCA preservatives such as Celcure A and Tanalith C have proved to be very reliable in service and there remain only minor criticisms of these systems. Firstly, the moisture content of the treated wood can fluctuate in service in the same way as for untreated wood and there is a tendency for checks and shakes to develop which may penetrate through the treatment if it is confined to the relatively permeable sapwood. This problem would be insignificant if treatment could be restricted to only permeable species such as Corsican pine or species such as Scots pine in which non-durable but easily treated sapwood is associated with relatively durable heartwood. However, in species such as spruce which combine impermeability with a lack of durability CCA treatment is unrealistic, as it fixes too rapidly and is unable to improve its zone of protection by diffusion so that splits may expose non-durable heartwood. Thirdly, some deterioration has been observed in hardwoods, even when treated at high retentions, apparently through poor microdistribution of the toxic components which do not fully protect the cell walls although they may be present in large amounts in the pores or vessels. Finally, there are criticisms of the toxicity of CCA preservatives, particularly in relation to their arsenic content, but these criticisms are common to other preservative systems and they will be discussed later.

There are now many copper–chromium–arsenic (CCA) wood preservatives, although they all originate from the three development routes that have been described. Within the CCA group the copper, chromium and arsenic components are present in varying proportions and incorporated variously as oxides or salts. To add to this confusion the standard specifications define these products in terms of an arbitrary selection of oxides or salts, whatever the actual compound that is incorporated in the individual formulation. Thus the copper content is expressed as $CuSO_4 \cdot 5H_2O$ in British Standard 4072 but as CuO in the American Wood Preservers' Association Standard P5, whatever the nature of the compound incorporated in the actual formulation. It might appear to be logical to describe preservatives in terms of their toxic elements alone, a system that has been used in Scandinavia and which certainly enables the ratios of the toxic elements to be readily compared, but it appears that if the active elements are described in terms of their equivalent oxide content, the total preservative activity of the formulation can be judged from its total active oxide content, thus enabling both the ratios of the active components to be compared and the probable preservative activity of an individual formulation to be assessed without difficulty.

For many years the American Wood Preservers' Association specifications defined copper–chromium–arsenic wood preservatives in terms of their equivalent salt contents as British Standard 4072 but this system was abandoned in 1969 and preservatives have since been defined in terms of their equivalent active oxide contents:

AWPA CCA	Type A	Type B	Type C	
Hexavalent chromium	65.5	35.3	47.5	CrO_3
Copper	18.1	19.6	18.5	CuO
Arsenic	16.4	45.1	34.0	As_2O_5

According to the specification the hexavalent chromium can be incorporated as potassium or sodium dichromate, or as chromium trioxide. The bivalent copper can be incorporated as cupric sulphate, basic carbonate, oxide or hydroxide. The pentavalent arsenic can be incorporated as sodium arsenate or pyroarsenate, arsenic acid or arsenic pentoxide. Mixtures of salts involve significant concentrations of inactive ions such as sodium and sulphate and these reduce the effective concentration of the preservative, which is expressed as a percentage indicating the active oxide content. For example, a BS 4072 type 1 preservative such as Celcure A has an active oxide content of 61.1% whereas a type 2 preservative such as Tanalith C has an active oxide content of 59.05%. There are various products that comply with this British Standard but, as they are not actually manufactured from the salts listed in the specification, their activity may not be 100%. Some have an effective activity of less than 100% but K33 paste would have an effective activity of 125% according to this British Standard if it complied with the standard in terms of the ratios of the active elements. In contrast, K33 paste has an active content of 75.4% according to the AWPA system as this is the concentration of active oxide actually present in the formulation, the inactive component consisting solely of water.

Greensalt – Langwood – Boliden CCA – Koppers CCA-B – Osmose K-33 – Crom–AR–Cu (CAC) – Wolman CCA – Wolmanac CCA

In the American specification, CCA type A represents the original Ascu type formulations and includes Greensalt and Langwood which possess a high chromium content. CCA type B was introduced solely for K33 type formulations and includes Boliden CCA, Koppers CCA-B and Osmose K-33. CCA type C includes products such as Crom–Ar–Cu (CAC), Wolman CCA and Wolmanac CCA are similar to BS 4072 preservatives. The American specification tolerates some variation from the nominal ratios defined above and this permits products conforming with both types 1 and 2 in the British specification to be formulated to meet also the American CCA type C specification.

It will be appreciated that considerable care must be taken in interpreting retention figures in different countries. A British Standard retention of 10 kg/m^3 (0.6 lb/ft^3) means that this retention must be achieved in terms of the salt mixture defined in the specification, a system that was originally used in America and which is still widely used in some other countries. This British Standard preservative has an active oxide content of about 60% so that a retention of 10 kg/m^3 is equivalent to a retention of 6 kg/m^3 in terms of the AWPA specification. In the Nordic countries and many other areas it is conventional to define retentions in terms of the preservative as supplied, so that a K33 retention of 10 kg/m^3, based on paste with an active oxide content of 75.4%, would be equivalent to 7.54 kg/m^3 in terms of the AWPA specification; K33 was also available as a dry powder product with a higher active oxide content, and as a solution with a K33 content of 60% and active oxide content of only about 45%. In order to avoid confusion it is recommended that all specifications should express retentions in terms of active oxide content as in the AWPA system as this also indicates the relative effectiveness of the product, and the contents should be clearly stated on all literature and labels.

Unfortunately, this system would not completely avoid confusion. In the British Standard retentions are expressed as overall retentions although only the sapwood is treated in species such as Scots pine in which the heartwood is relatively impermeable. This system has now been abandoned in the Nordic specifications, with overall retentions now replaced by sapwood retentions which can be realistically controlled and readily checked by analysis. European redwood or Scots pine from the Baltic area currently contains an average of about 50% sapwood so that a sapwood retention of 10 kg/m^3 is equivalent to an overall retention of about 5 kg/m^3, although it must be appreciated that the overall rtention will vary depending on the percentage of heartwood.

Tanalith C (Tancas C) – Tanalith CCA – Tanalith CT106 – Tanalith Plus

Tanalith C with an active oxide content of 59% is known as Tancas C in some areas such as Finland. Tanalith C (CT106) is a concentrated version of this product containing about 62% active oxides, but another concentrated product Tanalith CCA contains about 72% active oxide so that it can be used at the same retentions as K33. Tanalith Plus is the name applied when an emulsion additive is used in an attempt to improve the moisture resistance of the treated wood. The emulsion additive is prepared by dissolving waxes and a surfactant in a solvent, and it is then added at about 2% to a normal Tanalith C treatment solution. The emulsion tends to be unstable, through the low pH and the oxidizing properties of the hexavalent chromium in the preservative.

Celcure A – Celcure AP

Another interesting development was Celcure AP, a paste version of Celcure A which avoids toxic dust problems during solution preparation. This formulation is manufactured from cupric oxide, copper sulphate, sodium dichromate and arsenic pentoxide, a small amount of water being added to adjust the total active oxide content to 59.7%. The oxide contents are closely similar to those for Celcure A and this product therefore conforms with British Standard 4072 type 1 without the need for any concentration adjustment.

Celcure A – Celcure AN – Tanalith C – Tanalith CA – Tanalith NCA

The situation in New Zealand illustrates the further developments that are possible. The original Celcure copper–chromium (ACC) preservative was used at first but was found to be unreliable against insects and copper-tolerant fungi. Boric acid was added but, while this improved the insecticidal activity, it was significantly leachable and did not markedly improve the performance against the copper-tolerant fungi. This situation coincided with the problems in water-cooling towers in the United Kingdom, which resulted in the addition of arsenic and the development of Celcure A which was approved in New Zealand in 1959; the similar competitive product Tanalith C was introduced at about the same time. In 1961 the arsenic content in Tanalith C was increased to give Tanalith CA so that lower retentions could be used for the treatment of Radiata pine in buildings where the principal deterioration hazard was considered to be insect attack, Common Furniture beetle being the most widespread hazard but termites representing the most severe deterioration problem in some areas. In 1966 Tanalith NCA was introduced; in this the copper content was increased at the expense of the chromium content so that a lower retention could be used for the treatment of wood in ground contact and where there is also a risk of insect infestation. Celcure AN, a high-copper formulation similar to Tanalith NCA, was approved in 1967, although it was not introduced until 1969. Boliden S25 was also used in New Zealand, although it was replaced in 1967 by Boliden K33. These various CCA preservatives are compared in Table 4.1.

The main hazard in buildings is insect borer attack and it will be seen that the approved retentions therefore relate to the arsenic content, apparently designed to achieve a retention of about 0.9 kg/m^3 As$_2$O$_5$. For wood in ground-contact such as poles, piles and fence posts, the main risk is fungal decay and it will be seen that the approved retentions, which are based on experimental results in laboratory tests and long-term stake trials, relate to the total active oxide content of each formulation, although the Tanalith CA retention is slightly higher than might be expected from this relationship. The use of the high arsenic formulations cannot be recommended as the excess arsenic is not fixed and, when used at lower retentions where insect attack is the main normal hazard, the

TABLE 4.1 Comparison of CCA preservatives

	Active oxides Content % (ratio %)				Approved retentions kg/m^3 (lb/ft^3)	
	CuO	CrO_3	As_2O_5	Total	Ground	Building
Tanalith C	11.1 (18.8)	30.6 (51.9)	17.3 (29.3)	59.0	12.0 (0.75)	5.6 (0.35)
Tanalith CA	11.1 (15.2)	29.8 (40.7)	32.3 (44.1)	73.2	12.0 (0.75)	3.2 (0.20)
Tanalith NCA	14.8 (21.2)	24.2 (34.7)	30.7 (44.0)	69.7	10.2 (0.64)	3.2 (0.20)
Celcure A	10.2 (17.0)	27.2 (45.3)	22.7 (37.8)	60.1	11.2 (0.70)	5.0 (0.31)
Celcure AN	14.9 (22.2)	21.5 (32.0)	30.8 (45.8)	67.2	10.2 (0.64)	3.2 (0.20)
Boliden K33	14.8 (19.6)	26.6 (35.3)	34.0 (45.1)	75.4	8.5 (0.53)	3.2 (0.20)

fungicidal protection may not be adequate if accidental wetting should occur and there is then a danger that arsenic-resistant fungi may generate toxic arsine gas.

After-glow suppression – 3.S CCA – 3.O CCA – Celcure AG – Tanalith AG

Normal CCA preservatives used for fence post treatments in Australia have been found to suffer from after-glow or continuing charring following grass fires. The Government CSIRO laboratories have developed and patented a modification involving the addition of zinc oxide and phosphoric acid to a normal CCA formulation, to give type 3.S CCA based on salts, or type 3.O CCA based on oxides and containing 7.32% Cu, 10.94% Cr, 10.28% As, 3.93% Zn, and 4.61% P; Celcure AG and Tanalith AG are commercial formulations of this type. These modifications are not new – Gunn added zinc and phosphorus to the original Celcure (ACC) formulation to form Celcure F, the fire retardant version.

CCA and ACC fixation

There have been various explanations of the fixation of copper–chromium and copper–chromium–arsenic preservatives. In pine it is probable that some of the copper reacts with or condenses on cellulosic wood components, probably forming a copper–cellulose complex. The remaining copper reacts with dichromate to produce mixed copper chromates. Excess dichromate is reduced from the hexavalent to trivalent state and then reacts with any arsenic present or, if arsenic is absent, it is absorbed onto the wood. Arsenic is fixed principally by trivalent chromium, probably as $CrAsO_4$, although some arsenic may be absorbed onto the wood elements. At high arsenic concentrations some may be precipitated as copper arsenate. At low solution concentrations, absorption by the wood elements may be the major fixation process. The influence on fixation, of changes in the toxic component ratios, has been examined. The ratio of CrO_3 to As_2O_5 must exceed 1.5 to ensure arsenic fixation but if the ratio exceeds 2 the excess chromium is wasted. The ratio of CrO_3 to CuO should be at least 2. In order to ensure maximum fixation the toxic elements in a CCA preservative should be present at ratios of about 41–50% CrO_3, about 17% CuO and about 42–33% As_2O_5. These optimum ratios are met in CCA-C and BS 4072 formulations; as a result CCA-A formulations are declining in use and CCA-B formulations such as K33, with their excessive arsenic contents, have been withdrawn in most countries.

The copper–chromium–arsenic or CCA formulations are the most reliable general-purpose preservatives currently available, providing excellent protection against all types of fungal and wood-borer deterioration. The rapid fixation remains a problem as it limits the further diffusion of the preservative when it is used to treat relatively impermeable species and it is still frequently claimed that the Wolman FCAP preservatives, with their slower fixation, are more reliable in these circumstances. In fact, the FCAP preservatives are not so well fixed, and the fluoride components are slowly lost as hydrogen fluoride. Some of the formulations which were described earlier, and which rely on ammonia or acetic acid loss for fixation, can combine all the advantages of the ACC and CCA formulations, with deeper penetration in impermeable woods and more uniform microdistribution in hardwoods.

Arsenic toxicity

The arsenic content in CCA, FCAP, CAA, ZAA, ACA, ACZA and similar formulations is a serious problem. It is unfortunately true that cattle have been poisoned as a result of licking treated transmission poles and fence posts but this normally occurs only in areas where there is a natural salt deficiency and the danger can be completely avoided by providing proper salt licks and using only highly fixed preservatives. Arsenic preservatives are banned in buildings in some countries such as Finland, yet in others they are happily used even for the treatment of playground equipment. In Switzerland, arsenic preservatives are banned for the treatment of transmission poles as they may introduce environmental pollution. In most countries these hazards are considered to be insignificant with CCA preservatives in view of their excellent fixation (although fixation is only reliable with CCA-C and BS 4072 formulations), and the main fears are related to the possible volatilization of arsenic when treated wood is destroyed by burning.

Yet another fear is the danger of arsine poisoning. In about 1890 several fatalities occurred in homes and these were eventually attributed to arsine poisoning. Wallpaper, decorated with arsenical dyes, had been attacked by the fungus *Scopulariopsis brevicaulis*, at that time known as *Penicillium brevicaule*; it is now known that other fungi can generate arsine in this way. There is no danger of arsine poisoning from preserved wood provided that the fungicidal components are able to control the fungi that generate arsine. If preservation is required against insect attack alone, as for the treatment of timber in accordance with the Australian Quarantine Regulations or for Furniture Beetle control in New Zealand, there is clearly a temptation to use just a simple arsenical preservative, and there is then a danger of arsine poisoning if fungi are able to develop. Many insects are dependent on or encouraged by the presence of fungal attack, and fungicidal components in multicomponent formulations therefore assist in their control. This factor is completely ignored in the Australian and New Zealand situations where the minimum retentions of approved preservatives are based solely on an arsenic retention of about $0.97\,kg/m^3$ As_2O_5.

More recently there have been fears in the United States concerning the carcinogenic dangers associated with the arsenic contents in wood preservatives. There have been extensive enquiries and the current evidence suggests that the carcinogenic properties are largely associated with arsenic trioxide As_2O_3 and arsenites rather than with the arsenic pentoxide As_2O_5 and the arsenates that are used in modern preservative formulations. The carcinogenic dangers associated with arsenical wood preservatives are slight, but arsenic pentoxide is prepared from the trioxide and increasing controls on the latter compound are likely to introduce manufacturing difficulties, perhaps leading to scarcity and increased cost. The dangers associated with fixed-arsenic wood preservatives are often exaggerated, sometimes by manufactures of competitive products; the enquiries in the United States were prompted by the cement and concrete industry which feared the competition of pressure-treated wood foundations in domestic construction!

Arsenic dangers always attract the most attention but chromium may represent the greatest hazard in wood preservation. If properly formulated preservatives are used in a competent and responsible manner the dangers are very slight and this is confirmed by the very low level of illness or injury in this long-established industry but it must be recognized that dangers exist. Tropical conditions discourage the use of protective clothing and operatives then frequently suffer from chrome ulcers which are painful and difficult to heal, but it is interesting to note that there are no arsenic problems, and the injuries sustained are an indication of poor plant hygiene and control rather than serious criticism of the preservative formulations involved.

Copper–chromium–boron – Wolmanit CB (Ahic CB)

There have been various criticisms of copper–chromium–arsenic or CCA preservatives, particularly with regard to the possible dangers associated with their arsenic content and the problems that arise in the treatment of relatively impermeable woods through their rapid fixation. There have been various attempts to overcome these difficulties, the best known being the replacement of arsenic by boron to give a copper–chromium–boron or CCB formulation. Combinations of chromium and boron were first proposed by Wolman in 1913 and the early development of Celcure included copper formulations in which borates were used in place

of dichromates. It is often suggested that these ideas were first combined in the CCB preservative Ahic CB, subsequently renamed Wolmanit CB, which was first marketed in Germany in 1960, although the CCB preservatives were actually first developed by Kamesan in India during World War II. The toxic components in Wolmanit CB are at concentrations equivalent to 10.8% CuO, 26.4% CrO_3 and 25.5% H_3BO_3.

The development of CCB preservatives was severely criticized particularly by manufactures of CCA preservatives, as it was evident from laboratory experiments that the boron was largely unfixed, although this disadvantage is offset in actual service by the deeper diffusion of boron that is achieved in resistant heartwood – a very important factor where heartwood is non-durable, as in spruce. Trials on poles in service have given very good results and it has been found that CCB preservatives perform particularly well on impermeable species such as spruce; good penetration is achieved through the continuing diffusion of the borate but leaching in service is limited by the low permeability. However, the Nordic-recommended retentions based on stake trails suggest that on Scots pine the performance in ground contact depends on the cupric oxide and chromium trioxide contents alone, the borate content making no significant contribution in this test situation, although the test stakes are relatively small and the leaching of unfixed components is exaggerated compared with service performance in larger section poles.

Tanalith CBC – Celcure M

Other manufacturers produce CCB products such as Tanalith CBC and Celcure M, but they tend to be relatively expensive because of the high retentions that are demanded by some approval authorities, and they are usually promoted in markets such as Switzerland where there is resistance to the use of CCA preservatives because of the arsenic content. Slower fixation enables CCB preservatives to be applied by the Boucherie sap displacement method, and the development of Wolmanit CB stimulated considerable improvements in the design of the caps that are used for this process.

Celcure N – Tancas CC

As the boron content in CCB preservatives does not significantly contribute to ground contact performance, at least in the Nordic small-stake trials, the best technique to avoid the criticisms of CCA preservatives might be to omit the arsenic and return to the old copper–chromium (ACC) formulation now usually known as Celcure N. Tancas CC has been introduced in this way as an alternative to Tancas C (CCA), and such changes are certainly realistic in situations where there is no significant risk of insect borer attack or resistant fungi such as *Poria* species. The original Celcure formulation is still extensively used with complete success in the Netherlands, Sweden and parts of the United States; it is particularly suitable for agricultural purposes as it is completely free from the livestock poisoning dangers associated with preservatives containing arsenic.

Copper and zinc borates

Another obvious solution is to develop borate fixation systems. Ammoniacal copper borate systems were first proposed in about 1965 and were further developed (by the author's laboratory) in about 1978. They have never been extensively adopted, probably mainly for commercial reasons; although the ammonia liberated during the final vacuum stages of impregnation and during subsequent slow fixation is unpleasant and difficult to control, a problem associated with all preservative systems which fix by ammonia volatilization. An alternative simpler system (also developed in the author's laboratory) uses mixtures of copper or zinc with boric acid and fixes through the volatilization of a small amount of acetic acid; the ratios can be adjusted to form several alternative borates

but the best resistance to leaching is achieved using an excess of copper or zinc.

Boliden P50

Boliden P50 was developed in Sweden as an arsenic-free replacement for Boliden K33 with the arsenic pentoxide replaced partly by phosphorus pentoxide and partly by an increase in the cupric oxide content. The currently recommended Nordic retention suggests that its performance depends upon the total oxide content, including the phosphorus pentoxide. Whilst this preservative can be expected to perform well in normal ground-contact conditions it is not clear that it has any advantage over simple copper–chromium formulations; there must be doubts about its activity against borers and it is known that phosphates actually encourage the development of some stain fungi.

Cuprinol Tryck (KPN)

Cuprinol Tryck, originally Cuprinol KPN, was developed by Häger as an alternative to both CCA products and his earlier KP Cuprinol system which will be described later. This is an ammoniacal formulation; copper carbonate is dissolved in caprylic acid and ammonia is added to form cuprammonium caprylate.

Basilit CFK

Basilit CFK is another ammoniacal formulation in which the toxic elements are chromium, fluorine and copper as indicated in the name. This formulation consists of a mixture of very soluble copper hexafluorosilicate, ammonium dichromate and a small amount of diammonium hydrogen phosphate. The Nordic retention recommendations suggest that only the cupric oxide and chromium trioxide contents contribute to the preservative action in ground-contact conditions and any contribution from the fluoride is insignificant. This preservative thus possesses the same disadvantages as other copper and copper–chromium systems, such as low activity against insect borers and copper-resistant fungi, but it achieves more efficient penetration than the rapid-fixing copper–chromium (ACC) and copper–chromium–arsenic (CCA) systems.

BFCA salts

The BFCA salts developed in Australia contain boron, fluorine, chromium and arsenic. They were originally introduced in 1955 but the current formulation dates from about 1963. Wood is treated by immersion in concentrated solutions of the salt mixture, which can then penetrate by slow diffusion.

Diffusion treatments in plywood and poles

While referring to Australia it is worth mentioning several special diffusion techniques. Plywood can be protected against insect attack by adding sodium arsenite to the alkaline phenolic adhesive; sodium arsenate or arsenic pentoxide can be used, and may be easier to add as they are more soluble, but they are much more expensive. Adequate protection is obtained at a retention of 0.8 kg/m^3 As$_2$O$_3$ and veneers up to 2.5 mm are completely penetrated. However, this treatment cannot be recommended as it gives no significant protection against fungal decay and, if the plywood is wetted through any cause, there is a danger that arsenic-resistant fungi will develop which may generate toxic arsine gas. Copper hexafluorosilicate, the very soluble compound used in Basilit CFK, can be used at high concentrations for injection into poles as a remedial treatment to control progressive heart rot – the initial rot ensures that good distribution is achieved whilst the treatment will ensure that no further damage occurs. This compound is relatively inexpensive as sodium hexafluorosilicate is available in Australia as a waste from processing phosphate rock for the fertilizer industry.

Double diffusion

It will be appreciated from the earlier description of the decay of treated wood in cooling towers that this deterioration represented a serious and continuing problem. The Soft rot damage was initially superficial but progressively increased in depth on the relatively thin-section fill slats whilst, in the Celcure-treated towers, copper-tolerant fungi were attacking structural members and the mist eliminators. The Soft-rotted zone was very permeable and was used with considerable success in conjunction with a double diffusion remedial treatment process designed to halt further decay. The towers were operated in order to ensure that the fill was thoroughly wetted and it was then flood-sprayed with copper sulphate solution which diffused into the water in the Soft-rotted layer. After sufficient time had been allowed for deep diffusion to occur a second application, of sodium chromate, was made to precipitate the copper as copper chromate.

Another double-diffusion treatment that is worth mentioning and which was originally developed in the United States during World War II is currently in use in Papua New Guinea as it is a relatively simple treatment, suitable for use in developing countries. Dry wood is immersed in a 3% copper sulphate solution at about 88°C (190°F) for a period of seven hours and then the solution, still containing the wood, is allowed to cool for about sixteen hours. The wood is removed, any excess copper sulphate on the surface is rinsed away, and the wood is then immersed for 48 hours in a 10% sodium arsenate solution. The process was modified by using a mixture of sodium arsenate and dichromate for the second treatment to improve fixation. It is also possible to treat in the same way using sodium fluoride followed by copper sulphate.

Boron compounds

Perhaps the earliest record of the use of boron in wood preservation is in the chromium–boron preservative developed by Wolman in 1913. Twenty years later boron was proposed as a component in Celcure, principally as a replacement for dichromate in the flame-retardant formulation Celcure F. Borates possess fungicidal, insecticidal and flame-retardant properties, and their use in multicomponent preservatives such as CCB and the various copper and zinc borate systems has already been described. As borates combine both insecticidal and fungicidal properties they are efficient preservatives even when used alone, although they are not considered suitable for use when treated timber will be subjected to leaching or ground contact conditions. This restriction is based on the observation that the preferred borates are soluble, but it ignores the fact that borates can achieve exceptional penetration and, that when the sodium ions are neutralized by atmospheric carbon dioxide, the final boric acid deposit possesses very low solubility at normal temperatures.

Timbor

Borates can be applied by normal impregnation methods but their greatest value is in diffusion treatments. Freshly felled and converted green wood with a moisture content in excess of 50% is treated with the borate preservative by immersion or spray, and the treated wood is then close-stacked and wrapped or placed in storage rooms to prevent moisture evaporation, thus allowing the borate treatment to diffuse deeply. Boric acid and sodium tetraborate (borax) are insufficiently soluble but much higher concentrations can be achieved if a solution is prepared using 1 part boric acid to 1.54 parts sodium tetraborate decahydrate; this mixed solution is dried to produce Polybor, known as Timbor when used as a wood preservative and corresponding approximately to disodium octaborate tetrahydrate $Na_2B_8O_{13} \cdot 4H_2O$ which has a boron content equivalent to 117.3% H_3BO_3.

Diffusol

Despite this high boric acid equivalent and excellent solubility it is still necessary to heat the solutions to maintain the required concentrations, which vary with the thickness of the

wood to be treated, so that the retention is uniform whatever the cross-section area. Thus in 25 mm (1 in) thick wood a minimum solution concentration of 20% H_3BO_3 is required and a minimum temperature of 40°C (104°F) is therefore necessary, whilst at 75 mm (3 in) thickness the required concentration increases to 40% and the temperature to 57°C (135°F). Timbor diffusion has been widely used throughout the world but the minimum storage period of 4 weeks per 25 mm (1 in) thickness necessarily involves considerable capital cost and high interest rates have discouraged the use of the system in many countries, although it remains extremely attractive for the treatment of non-durable tropical hardwoods in developing countries. Diffusol is a thickened borate treatment which can achieve adequate surface loadings without heating.

Timbor rods – Boracol 20 – Boracol 40 – Trimethyl borate (TMB)

Sodium octaborate tetrahydrate is also available as Timbor rods which can be inserted in drilled holes in wood at risk such as window frames, joist and beam ends in damp external walls, and even poles and sleepers (ties), producing a preservative solution if the wood becomes wet. Boracol 20 and 40 are concentrated borate solutions containing 20 and 40% disodium octaborate tetrahydrate respectively; they are generally used as alternatives to Timbor rods by injection into drillings in wood components subject to severe fungal decay hazard, usually as remedial treatments. Trimethyl borate (TMB) is a very volatile compound which can be applied as a vapour-phase treatment to achieve deep penetration in woods which are impermeable to normal liquid treatments; subsequent steaming hydrolyses the TMB to deposit boric acid in the wood.

Borester 7

Borates are particularly useful as treatments for hardwoods that are susceptible to Lyctid beetle attack as they are effective at very low concentrations, applied by immersion or spray; boric acid solution is frequently used for this purpose in Australia. Borates are also extremely effective in the control of stain fungi, although they are most effective as alkaline high-pH treatments so that sodium tetraborate is more reliable than boric acid or the highly soluble mixture. However, they are relatively inefficient against superficial moulds such as *Penicillium* and *Trichoderma* species so that they must be used in combination with other toxicants, such as sodium pentachlorophenate, as described later in this chapter. Borate esters enable boric acid to be used in organic-solvent formulations; hexylene glycol biborate, Borester 7, is most extensively used in this way.

Chlorophenates

Sodium pentachlorophenate is well known alone or in combination with borates as a sapstain control treatment but it can also be used as a component in wood preservative formulations. Its use as an alternative to dinitrophenol in Wolman FCAP salts has already been mentioned but it also forms the basis of several two-stage treatments in which copper or zinc pentachlorophenates are deposited. For example, copper sulphate and sodium pentachlorophenate solutions can be added to fibre-board to precipitate copper pentachlorophenate, an organometallic compound which will be considered later and which is also the active component in KP-Cuprinol and several other preservatives.

There are hundreds of different water-borne simple salt and multicomponent preservatives currently on the market throughout the world or which are historically significant. It is impossible to mention them all in a brief description and it would, indeed, serve little purpose as there are continuous changes. This description is therefore restricted to a few of the most important formulations in an attempt to define the general principles involved. Multicomponent water-borne preservatives are reliable and eco-

nomic. They can achieve excellent fixation and provide permanent, clean and safe treatment of wood. Their main disadvantages are their failure to control changes in the moisture content in wood so that there is a danger that checks and shakes may develop through the preserved zone if the penetration is limited. Some of the components used, particularly arsenic, are very toxic but excellent fixation ensures that treated wood is entirely safe with properly designed formulations. There are several other minor criticisms of water-borne preservatives which should be mentioned. The electrical insulation value of treated wood is important for transmission poles and railway sleepers (ties). With CCA preservatives wood has a relatively low electrical resistance when freshly treated but this increases steadily with drying; the resistance is never as high as with creosote treatment but with oxide preservatives it is generally the same as for untreated wood, although with salt preservatives it may be less and perhaps too low for track signalling systems in some sleeper (tie) treatments.

4.4 Organic compounds

Historically, the most important organic preservatives were the fractions obtained by the distillation of tar, principally from coal; preservatives of this type have already been described in detail in Section 4.2. It was observed that changes in the composition of these distillates affected preservative performance, apparently through changes in the concentrations of individual components. At first, changes in composition were caused mainly by processing, to remove compounds that were useful for other purposes, but there were later attempts to control processing to achieve the most reliable creosote wood preservative. There were subsequently attempts to fortify tar-oil products, either by processing such as by the chlorination used in Carbolineum Avenarius, developed about 1888, or by the addition of toxicants, such as the copper salts added by Nordlinger about 1900, and the arsenic compounds that are today added to creosote in Australia to enhance insecticidal properties. Attempts to replace creosote with petroleum distillates of a similar physical nature were unsuccessful as it was soon discovered that they lacked preservative properties, and toxicants needed to be added; pentachlorophenol in heavy oil has been extensively used in the United States as an alternative to creosote.

The addition of toxicants to heavy petroleum oil was originally an attempt to find alternatives to creosote at times of scarcity. While the new formulations were designed to possess all the advantages of creosote in terms of toxicity and even the limited volatility that is so valuable in stabilizing the moisture content in treated wood, they also retained the disadvantages such as the dirty appearance and the tendency to bleed, largely as a result of the selection of a comparatively crude and inexpensive solvent. True organic wood preservatives consist of a solution of toxicants in volatile solvents and give perfectly clean treatment if the solvent is completely volatile and the toxicant is colourless. The preservative action depends solely on the persistent toxic deposit and the solvent has no preservative action, however expensive it may be. The use of such solvents can be justified only when water-borne preservatives are unacceptable. The main problem with the use of water is the swelling that it causes as this is unacceptable for worked joinery (millwork); even if the treated wood is dried as there is still a danger of permanent distortion. For such purposes the extra cost of an organic solvent can be justified and worked joinery now constitutes the main market for organic solvent-based preservatives.

Organic solvents – proprietary advantages

Whilst the organic solvent may not contribute directly to the preservative properties it is certainly of considerable importance. Non-polar

light petroleum distillates have low viscosities and are able to penetrate rapidly into dry wood so that they are particularly suitable for use in preservative formulations that are designed for superficial application by brush, spray or immersion. However, it is not sufficient to achieve deep penetration by the preservative formulation as the subsequent volatilization of the solvent may cause the toxicants to return to the surface where they will be particularly susceptible to losses by volatilization and leaching. Co-solvents are frequently used, sometimes to speed up the solution of the toxicants during manufacture, but often to provide high-viscosity residues or tails which will retain the toxicants whilst the carrier solvent volatilizes, ensuring proper distribution. In some cases the use of co-solvents distinctly reduces the apparent toxicity of an active component, perhaps due to deeper distribution or a protective action which also ensures greater persistence and a longer effective life. Whilst many products may be based on the same concentration of a particular toxicant, their performance may vary widely and it must not be assumed that apparently similar proprietary products achieve similar performance.

Nitrated compounds

Nitration may increase the fungicidal activity of compounds such as phenol, cresol, xylenol, naphthol and anthranol. Various nitrated compounds were proposed as wood preservatives in the 19th century but few were adopted commercially as tests indicated that high retentions of, for example, dinitrophenol or dinitro-o-cresol were required when they were used alone.

Raco – Antinonnin – Antingermin – Mykantin

Raco consisted principally of nitrated phenols whilst Antinonnin, which was introduced in 1892 but still in use in 1913, consisted of a mixture of potassium dinitro-o-cresolate, soft soap and water. In Antingermin copper was used in combination with dinitro-o-cresol. However, the most important use of nitrated organic compounds was in combination with salts in the Wolman-type preservatives as described in Section 4.3. Whilst there have been many recent attempts to develop in-situ ground-line treatments to extend the life of treated poles, it is interesting to note that a sodium dinitrophenate formulation known as Mykantin paste was developed by Falck for this purpose as long ago as 1912.

Chlorinated compounds – chlorophenols

There were many attempts to enhance the preservative activity of creosote by chlorination, stimulating interest in chlorinated phenols, although they were not seriously considered as wood preservatives until much later when Hatfield in the United States completed an assessment of their activity. He first investigated tetrachlorophenol and later pentachlorophenol, whereas Iwanowski and Turski at the same time investigated di- and trichlorophenols. By 1935 Hatfield was able to review the properties of the complete range of chlorophenols and chloro-o-phenylphenols, as well as their sodium salts. It was found that higher chlorination generally increased the fungicidal activity but also the melting point, giving more active and more persistent preservatives. Following completion of these studies, pentachlorophenol was introduced for exterior joinery (millwork) preservation in about 1936, and within a few years a 5% solution of pentachlorophenol was widely accepted as the standard organic-solvent formulation.

Pentachlorophenol solutions have been prepared using a wide range of solvents. A heavy persistent solvent is often selected where the preserved wood will be exposed to ground contact or severe weathering, as this will protect the pentachlorophenol from leaching and will tend to stabilize the wood as with a creosote treatment. Volatile solvents are employed when the preservative is intended for internal use or where the treatment must be paintable. Generally, co-solvents must be added, perhaps in order to obtain the necessary solvency power but also as anti-blooming agents to prevent the migration of the pentachlorophenol to the surface as the light carrier-solvent volatilizes. These anti-blooming agents are normally non-volatile such as dibutyl phthalate or trixylyl phosphate but paintability is improved if solid co-solvents are used such as rosin esters. The non-volatile co-solvent content is particularly important as it has a profound influence over the life of the preservation treatment as it largely determines the distribution of the pentachlorophenol and perhaps physically protects it from volatilization.

Phenylphenol

The water-soluble sodium, and occasionally potassium, salts of the chlorophenols are also extensively used, principally as sapstain control treatments for freshly felled green wood but also for masonry sterilization associated with remedial treatment against Dry rot. Sodium pentachlorophenate dust or solution spray is distinctly irritant and there have been several attempts to develop more pleasant alternatives. Sodium o-phenylphenate has largely replaced sodium pentachlorophenate in remedial treatment although it is little used in sapstain control; it is more expensive but less effective than sodium pentachlorophenate. Phenylphenol was originally a by-product of the production of phenol by the Dow process which has now been replaced by the cumene process, except in Poland. Some o-phenylphenol is synthesized in the United Kingdom but Poland is now the only country in which this compound remains available at a realistic cost that enables it to be used in stain control.

TC oil

While referring to phenol production it is perhaps worth mentioning that phenol itself is too soluble and too volatile to be used in wood preservation but the cumene process generates a phenolic residue, known in the United Kingdom as TC oil, which is very similar to creosote in its physical properties. It is a particularly useful preservative for exterior timbers such as fencing, as it has excellent colour retention properties. It is also free from the carcinogenic polycyclic hydrocarbons that limit the use of creosote. It is therefore surprising that these phenolic residues have not been proposed as wood preservatives in other countries where phenol is produced by the cumene method.

Although sodium pentachlorophenate is extensively used for sapstain control treatment, pentachlorophenol is relatively inefficient. This difference can be attributed to the high pH of the sodium pentachlorophenate solution which is alone sufficient to inhibit stain and can achieve complete control in the presence of limited amounts of toxicant. The addition to sodium pentachlorophenate of a buffer such as sodium carbonate, to maintain the high pH, considerably enhances the apparent stain control activity. A suitable phosphate can function as a buffer in the same way but the residual phosphate will encourage the development of surface mould. A borate buffer is therefore preferred as it also contributes to the toxicity of the formulation; borate alone is sufficient to control the stain and the sodium pentachlorophenate is needed only to control superficial moulds which are resistant to boron.

Pentabor

One of the most effective formulations consists of 1 part sodium pentachlorophenate with 3 parts sodium tetraborate decahydrate (borax) which can be used at the same concentration as the sodium pentachlorophenate alone. This formulation generally achieves improved stain control at lower cost and considerably reduces any dangers associated with the sodium pentachlorophenate, since this is reduced to only a quarter of the total content. Pentabor S is a formulation of this type but with half the water of crystallization removed, to concentrate the product and reduce transport costs. This ratio of the toxic components performs well for treatments in temperate areas but the proportion of sodium pentachlorophenate must be increased for the treatment of tropical hardwoods.

Chlorophenol toxicity

There are many proprietory wood preservatives containing chlorophenols and it is unrealistic to list them. Pentachlorophenol is the most important organic compound used in wood preservation. In the United States the wood preservation industry still uses about 20 000 tonnes of pentachlorophenol annually, despite concern regarding possible health hazards; whilst it is certainly true that chlorophenols are toxic, this is a problem that applies to all wood preservatives and they should therefore be handled with care. Some fatalities have occurred but these have all been associated with abnormal absorptions of pentachlorophenol, through a failure to take normal precautions. There are also fears that the dioxin impurities in chlorophenols may be particularly hazardous but, whilst there may be distinct dangers associated with 2,4,5-T, the well known herbicide which is a trichlorophenyl acetate, there is no evidence of similar dangers associated with pentachlorophenol or even the trichlorophenol with which it is sometimes contaminated – apparently because the latter is 2,4,6-trichlorophenol and generates a different range of dioxin impurities which are actually less toxic than the chlorophenols from which they are derived. Environmental theoreticians have suggested that pentachlorophenol should be replaced in stain control by less persistent trichlorophenol but this compound is less effective and must be used at higher concentrations, and the dioxin impurities tend are more toxic than those associated with pentachlorophenol. The continued use of pentachlorophenol would appear to be best in terms of both handling safety and environmental protection. Tetrachlorophenol is also sometimes used and represents an intermediate risk.

Pentachlorophenol (PCP) – pentachlorophenyl esters – Mystox LPL – Fungamin

Pentachlorophenol is essentially a fungicidal wood preservative possessing only limited activity against borers, principally those that are dependent on previous fungal decay or the presence of an intestinal flora. It is therefore frequently formulated with insecticides, particularly the contact insecticides that are described later in this chapter. It has no appreciable fixation to wood and is essentially a simple toxic deposit which normally possesses reasonable life through its relatively low volatility coupled with fairly deep penetration. Pentachlorophenyl laurate (Mystox LPL) and rosin amine-D pentachlorophenate (Fungamin) are compounds in which the pentachlorophenol has been reacted with other organic groups to increase the molecular weight and reduce volatility but solubilization is also assisted and the need for special solvents to prevent surface blooming is avoided. Copper and zinc pentachlorophenate are also useful preservatives; these are described in detail later in this chapter.

Cumullit – Parol

p-chloro-*m*-cresol was first proposed as a wood preservative in 1913 and was the main toxicant in Cumullit, first marketed in Germany in about 1917. This product performed well when applied by normal pressure impregnation methods but low retentions from brush treatment were inadequate; it is not known whether this was due to inadequate solution concentration. Parol was the potassium salt and was used for the treatment of transmission poles in Germany in World War I; in world War II it was again used either as the sodium salt or dissolved in ethanol before dilution to a concentration of about 1%. *p*-chloro-*m*-cresol is a fungicide and not significantly insecticidal. The compound does not appear to be currently used but it is mentioned here, as are several other compounds, as a potential wood preservative for use when established compounds become scarce or too expensive.

Chloronaphthalenes

The chloronaphthalenes were proposed as wood preservatives following the successful chlorination of creosote and the development of Carbolineum Avenarius. Individual chloronaphthalenes are rarely used commercially and mixtures resulting from the chlorination of naphthalene are usually described in terms of their melting point which increases with the degree of chlorination. Thus the mono- and dichloronaphthalenes are liquids whereas tri- and tetrachloronaphthalenes are solid waxes.

The mono- and di- compounds are distinctly fungicidal but this activity appears to be associated in part with their volatility and the fungicidal properties are reduced in the tri- and tetra-compounds, despite the higher chlorination which would normally suggest greater fungicidal activity. The explanation probably lies in their manner of use and evaluation – these compounds are used in superficial brush and spray treatments where the deeper distribution of the lower viscosity compounds accounts for their apparent fungicidal activity, whereas in an impregnation treatment the solid compounds combine greater protection against fungi with much greater permanence through their resistance to volatilization.

The insecticidal properties increase steadily with the degree of chlorination, whether the compounds are applied by impregnation or superficial application, and this fact is reflected in, for example, the South African standard which requires principally tetrachloronaphthalene to be used and thus defines a minimum chlorine content of 47% and a minimum softening point of 90°C (194°F).

The volatility of the mono- and dichloronaphthalenes results in a characteristic odour which is a distinct disadvantage. As the chloronaphthalenes are largely used in preservatives for superficial application their performance depends largely on their penetration. There have been several suggestions that this can be improved by the addition of various resins and waxes, and particularly by adding stearic or palmitic acids or esters; the latter act as surfactants, condensing onto the hydrophilic wood components and permitting the hydrophobic solution to penetrate. Surfactant systems of this type are extensively used as additives to road tar and bitumen to assist wetting of damp crushed stone.

Halowax – Wykamol (Anabol)

Halowax, consisting mainly of trichloronaphthalene gave excellent preservation against termite attack when evaluated in Panama in 1913, but the advantages of this material do not appear to have been immediately appreciated commercially. Chloronaphthalene wax was the principal toxicant in Anabol, a remedial-treatment wood preservative introduced in England in 1934 and subsequently renamed Wykamol. The excellent termite resistance of chloronaphthalene wax was also established in South Africa in about 1950; a mixture of 3.5% tetrachloronaphthalene as an insecticide with 2.0% pentachlorophenol as a fungicide in organic solvent was approved for use by pressure impregnation to meet regulations that required all wood in certain areas to be preserved against termites and Longhorn beetles.

Xylamon – Olimith C20 – Ridsol

Mono- and dichloronaphthalene were first proposed as wood preservatives in about 1920 and were the principal toxicants in Xylamon which was introduced in 1923. Similar products have been introduced more recently such as Olimith C20 and Ridsol in the Netherlands. Whilst the smell of these compounds is a disadvantage, the toxic vapour diffuses deeply after superficial treatment and these products are therefore particularly suitable for the eradica-

tion of House Longhorn beetle and other borers in remedial treatments.

Chlorobenzenes – Rentokil

Whilst referring to remedial treatments it should be added that, whilst chloronaphthalene wax was the principal toxicant in the product Anabol, later renamed Wykamol, this formulation also originally contained o-dichlorobenzene to give a deeply penetrating insecticidal vapour action. This component was replaced in Wykamol in 1939 by Rotenone and, after World War II, by Lindane contact insecticide, although it continued in use in other similar formulations such as Rentokil. o-dichlorobenzene (ODB) is a liquid whereas p-dichlorobenzene (PDB) is a solid but more volatile. The eradicant activity of PDB against Common Furniture beetle was first established in about 1915 but the excessive volatility results in a powerful eradicant action with only limited persistence. ODB is slightly less active but penetrates well into wood and gives excellent insecticidal preservative action coupled with good persistence. The fungicidal activity of the chlorobenzenes increases with the degree of chlorination as might be expected and the trichlorobenzenes have been proposed as wood preservatives. Hexachlorobenzenes have also been considered but their low volatility and low water solubility substantially reduce their effectiveness when applied as superficial preservatives, although they are probably much more effective when used for impregnation.

Gamma hexachlorocyclohexane – Lindane Gammexane

Hexachlorobenzene must not be confused with benzenehexachloride, a rather misleading name that was used in the past for hexachlorocyclohexane. The gamma isomer of haxachlorocyclohexane has been known variously as γ-BHC, Gammexane, γ-HCH and Lindane, although the two latter terms are now preferred. Lindane has been extensively used as a contact action insecticide in formulated organic-solvent preservatives, particularly for superficial application in remedial treatments and usually at a concentration of 0.3–0.5%. Activity decreases progressively through volatile losses that are initially very rapid but if reasonable penetration is achieved the treatment will give protection against egg-laying or emergence of adults following pupation for many years. This activity is affected by the presence of other components, especially some waxes, resins and non-volatile oils, which trap the Lindane, usually giving a lower initial toxicity but greater persistence; the value of different proprietory products may vary greatly, despite a similar Lindane content, through the presence of these apparently non-functional components.

Rotenone – Dicophane (DDT)

It is often suggested in literature on wood preservatives that Lindane and other contact insecticides were first considered as insecticidal components in about 1947–1953 but, whilst this was apparently the situation in Germany and perhaps the United States, Lindane was already in use in the United Kingdom in 1945 where it had replaced Rotenone, a natural contact insecticide derived from the Derris root, in Wykamol. Prior to the introduction of Lindane, Dicophane or DDT (dichlorodiphenyltrichloroethane) and the less effective contact insecticide 666 had also been considered but were less effective. DDT must be used at 2%, or even 5% where there is a severe insect hazard such as a risk of termite attack, to achieve the same protection as Lindane at about 0.5%, the insecticidal activity depending on the *pp*-DDT content.

Cyclodiene insecticides – Heptachlor (Chlordane) – Dieldrin (HEOD) – Aldrin (HHDN) – Endrin

The cyclodiene insecticides Chlordane (Heptachlor), Aldrin (HHDN) and Dieldrin (HEOD) have all been widely employed as soil poisons in termite control treatments but Dieldrin was

preferred as a wood preservative because of its lower vapour pressure and better persistence. This persistence has been a considerable disadvantage when Dieldrin has been used in agriculture and horticulture as the insecticide has accumulated in natural food chains, causing considerable infertility where these chains end in, for example, birds of prey and causing concern that mankind may be affected in a similar way. Endrin, a similar insecticide, is less persistent than Dieldrin and has therefore been more widely used in agriculture and horticulture but it is not so active as an insecticidal wood preservative. Whilst Dieldrin is not popular in wood preservation because of the bad publicity associated with these agricultural problems, none of the chlorinated hydrocarbon contact insecticides represent significant environmental risks when employed in wood preservation as these products control insects only within the confined environment of the wood so that they are unable to enter the natural food chains.

Whilst it may be an advantage to develop non-persistent insecticides for use in agriculture and horticulture it is clearly essential that extremely persistent insecticides should be developed for use in wood preservation. In most countries these contact insecticides continue in use as wood preservatives, perhaps subject to special licensing if their use is forbidden for other purposes, although generally Lindane is now preferred to Dieldrin. One difficulty is the limited availability of these insecticides now that they are no longer manufactured for agriculture, since it does not appear to have been generally appreciated that the wood-preservation market for these compounds is larger in many countries than for agriculture. Chlordane (Heptachlor) has been recently considered an as alternative to Dieldrin as it is less volatile and thus more persistent that Lindane, but there is some current concern regarding possible carcinogenic dangers associated with this compound.

Among the chlorinated hydrocarbon contact

insecticides Dieldrin, Aldrin and Chlordane are most persistent. Lindane is less persistent and DDT is not lost too rapidly but it is much less active. Marine exposure tests have shown that these contact insecticides possess excellent activity against the crustacean borers such as gribble, but little activity against the molluscan shipworms which are more susceptible to fungicidal preservatives. It has been found that a chlorinated hydrocarbon contact insecticide such as Dieldrin or Lindane in creosote provides excellent preservation in marine situations.

Organophosphorus and carbamate insecticides

The problems with chlorinated hydrocarbon insecticides have prompted the development of many alternatives, mainly for agricultural and horticultural uses, where only limited persistence is considered to be advantageous, so that most have been unsuitable for wood preservation where persistence is essential. Phenthoate, Fenitrothion, Malathion, Dichlorvos, Carbaryl, Diazinon, Fenthion, Arprocarb, Bromophos and Fenchlorphos have been considered, as well as many other organophosphorus and carbamate insecticides. The most promising were organophosphorus compounds; some of them such as Malathion were unsuitable due to their persistent and unpleasant odour but Fenitrothion and particularly Phenthoate were active and acceptable alternatives to Dieldrin and Lindane, although all organophosphorus insecticides are very toxic to humans.

Pyrethroids

The pyrethroids are much safer. They were originally derived from the pyrethrum daisy but the natural extracts are not persistent on wood although they have a very powerful knockdown insecticidal action which is valuable in flying-insect control. Synthetic pyrethroids have been developed with a range of properties and several of these compounds are now used in forest and mill treatments against pinhole and lyctid attack, in remedial treatments, and even in preservation treatments against termite and House Longhorn beetle attack. Decamethrin has rather high mammalian toxicity but Deltamethrin, Cypermethrin and particularly Permethrin have low toxicity and are now extensively used; they can give both eradication and long-term protection against wood-boring beetles, termites and crustacean marine borers such as gribble.

Many organic fungicides have been developed since World War II and some have been used in wood preservation. Most of these compounds originate from agricultural and horticultural research programmes aimed at developing fungicides with low persistence and it is not therefore surprising that very few of these new compounds have proved effective as wood preservatives, although some of them are used in stain control. Modern performance and safety approval schemes involve substantial cost so that developments for wood preservation alone are unrealistic as expensive new products cannot compete with established products, and only a few compounds developed since World War II justify mention here. Suspensions of Benomyl (Benlate) and Captafol (Difolatan) were introduced as alternatives to chlorophenols but they are not generally very effective; they leave only a superficial deposit which, whilst it may control surface stain and mould, has no influence over deep stain within the wood, and these fungicides are only reliable when formulated with borates which will control deep sapstain. Carbamates such as IPBC (Polyphase), chlorothalonil (Tuffgard, Tuffbrite), various isothiazolone (ITA) compounds, thiazoles such as MBT and TCMTB, and triazoles such as cebuconazole, propiconazole and azaconazole (Madurox) have performed well in stain control trials alone or in various formulations, and are also claimed to be suitable for wood preservation use but they are never as persistent and reliable as inorganic, organometal or even quaternery ammonium systems. The trihalomethylthio- compounds must be mentioned as they have been particularly to control stain in service beneath paint and varnish coatings. Captan is probably the least efficient of this group but this may be only a solubility factor as Folpet (Fungitrol 11) with the same trichloromethylthio- radical is very active. The dichlorofluoro- compounds Fluorfolpet (Preventol A3) and Dichlofluanid (Preventol A4) are also very effective and are preferred by many manufacturers as they are more readily soluble in most organic solvents whereas Captan and Folpet are virtually insoluble and can usually be applied only in relatively high viscosity systems, such as pigmented coatings or formulations with a relatively high resin content.

Various metal soaps and organometal compounds are used in organic solvent systems but they will be considered separately in the next section of this Chapter, together with organonitrogen compounds such as quaternery ammonium compounds.

4.5 Organometal compounds

Various organometal compounds have been used in wood preservation, particularly mercury and tin compounds. Metal soaps, particularly copper and zinc naphthenates, have also been used and, whilst they are metal esters rather than organometal compounds, they are considered to be more appropriate to this section. Similarly, the organonitrogen compounds are not organometal but they are included in this section as they are similar in many respects to organotin compounds.

Copper and zinc soaps

The preservative activity of metal soaps can be attributed principally to the metal content and this is now recognized in most specifications. Following treatment, the soaps hydrolyse and the acid is slowly lost by volatilization, leaving only the metal. Whereas the acid has an important eradicant function when metal soaps are used for remedial treatment, care must be taken to condition test blocks to ensure that most of the acid has dispersed when assessing preservative activity, to avoid enhanced performance which will be absent in actual service. Copper and zinc are effective fungicides but if they are applied at inadequate retentions they tend to be detoxified by tolerant fungi such as *Poria* species, and this can be clearly seen when inadequate loadings of green copper naphthenate are used, as fungal infections may decolourize the treated wood. Other metal naphthenates have been employed, such as calcium and barium, but the activity of these compounds must be attributed to the acid alone which is ultimately lost by volatilization.

Cuprinol

Metal naphthenates were first proposed as wood preservatives by von Wolniewicz in Russia in 1889 but they were first marketed commercially as Cuprinol in Denmark in 1911 and were introduced from there into Sweden in about 1920 and into England in 1933. Copper naphthenate is produced either by the fusion method in which copper oxide or carbonate is dissolved in heated naphthenic acid or by the precipitation method following double decomposition when sodium naphthenate is mixed with copper sulphate in aqueous solution. The very distinctive green copper naphthenate is an important organic-solvent wood preservative and clear zinc naphthenate is also extensively used. Both these soaps possess low insecticidal activity, except when free acid is still present, but adequate retentions give good fungicidal protection. Failures of treated wood in service are usually associated with the development of resistant fungi on wood treated by superficial brush or spray application.

Oborex Cu and Zn

Copper and zinc naphthenates are marketed throughout the world under a variety of names, such as Oborex Cu and Zn in the Netherlands, but it is unrealistic to list them – there were at one time 23 copper naphthenate and a further eight zinc naphthenate formulations registered in Sweden alone! In the United Kingdom the production of copper naphthenate increased considerably during World War II, largely as a result of the demand for preserved ammunition boxes and other military packaging which would not deteriorate in service, but also for treatment of canvas. Copper naphthenate has not been used in heavy petroleum oil to the same extent as pentachlorophenol, although such formulations inhibit the volatilization of the naphthenic acid and prolong the acid activity.

Acypetacs copper and zinc

Naphthenic acid has not been readily available in recent years and various synthetic acids have been considered as alternatives. Many organic acids are suitable but octanoic and versatic acids have been preferred in some countries, probably because they have been used as alternatives to naphthenic acid in soaps used as paint driers. Cuprinol in England has been particularly active in the search for advantageous acids, eventually adopting a mixture of linear and branched chain saturated aliphatic carboxylic acids derived from petroleum, the resulting soaps being described as acypetacs copper or zinc.

Copper and zinc pentachlorophenates

The formation of copper pentachlorophenate by precipitation from mixtures of copper salts and pentachlorophenates will be described shortly but copper and zinc soaps are also mixed with pentachlorophenol in organic-solvent solutions in the hope that the metal pentachlorophenates will be formed following the volatilization of naphthenic acid after treatment. The unpredictability of the reactions is clearly apparent with the copper formulations as they often retain the green colour of the copper naphthenate whereas copper pentachlorophenate is dark red. However, even in acid conditions, where the formation of pentachlorophenate is prevented

in this way, these mixed preservatives have considerable advantages over the use of their individual components alone as they possess a broader spectrum of activity and greater persistence, and numerous clear organic-solvent preservatives are therefore based on mixtures of zinc naphthenate and pentachlorophenol.

These preservatives are often prepared by adding zinc naphthenate to a normal pentachlorophenol solution containing the usual co-solvents and anti-blooming agents, yet if the toxicants are present at an appropriate ratio the zinc naphthenate content alone is sufficient to carry the pentachlorophenol without the need for these additives, an observation that is put to good effect in the South African Standard for such mixtures, which requires 0.31% zinc and 2.5% pentachlorophenol in normal preservative solutions applied by pressure impregnation. Naphthenates are also similarly used in conjunction with other anionic fungicides such as o-phenylphenol but many preservatives also contain trihalomethylthio type fungicides to improve the resistance to staining fungi.

KP-Cuprinol – Penta-Tetra-Copper

KP salt, which is known in Sweden as KP-Cuprinol, is another invention of Häger who was responsible for K33 and the earlier Boliden products BIS, S and S25. KP salt was first introduced in Sweden in 1955 and was supplied as two components – the K salt contained copper as the ammoniacal carbonate whilst the P salt contained sodium chlorophenate, usually tetrachlorophenate. The two salts were dissolved separately and then mixed to prepare a solution with a pH of 8.0–8.5 for full-cell impregnation containing 0.3% copper and 0.15% chlorophenol. After treatment the ammonia evaporated, reducing the pH and causing the precipitation of copper pentachlorophenate. As the fixation process occurred only after the treatment cycle was completed, KP salt could be applied by empty-cell processes, although the Lowry process was normally preferred. KP salt was applied in this way at the commencement of the Royal process for the treatment of joinery (millwork) in which impregnation is followed by heating under oil to remove the water and ammonia so that, when the wood is removed from the treatment vessel, it is also deeply impregnated with the oil. Penta-Tetra-Copper is a similar product, forming a mixture of copper penta- and tetrachlorophenate.

Cuprinol Tryck (Cuprinol KPN)

Cuprinol Tryck, originally Cuprinol KPN, was developed by Häger as an alternative to both CCA products and his earlier KP Cuprinol system. It is an ammoniacal formulation comprising copper carbonate dissolved in caprylic acid and ammonia added to form cuprammonium caprylate, the eventual volatilization of ammonia precipating cupric caprylate.

Copper 8-hydroxyquinolinolate (oxine copper) – Cunilate – Nytek GD – Cupristat

Copper 8-hydroxyquinolinolate, also known as oxine copper, has been introduced in recent years as an alternative to copper naphthenate. This compound does not present the distinctive naphthenic acid odour of copper naphthenate and it is also resistant to hydrolysis so that it can be solubilized in water; Cunilate 2174 is 10% of this compound solubilized in water. Wood preservatives of this type such as Nytek GD and Cupristat are also used for stain control. They have very low mammalian toxicity and are virtually free from tainting problems so that they can be used for the treatment of food packaging such as fruit boxes, sometimes with the addition of water-repellent components.

Organomercury compounds

The use of mercury compounds in wood preservation has already been described in an earlier section of this chapter and it will therefore be appreciated that mercuric chloride (corrosive sublimate) was an important wood preservative in the late 19th century. In 1910 mercury chlorophenate was developed to avoid the corrosive

properties of mercuric chloride but it was found to be comparatively uneconomic and was never used commercially. Various organomercury compounds were patented in Germany in 1926 but were not used significantly in Europe, although ethyl mercury and phenyl mercury compounds were introduced in North America in about 1930 and were soon extensively used in stain-control treatments. Ethyl mercury chloride, sulphate, phosphate and acetate have all been used, as well as phenyl mercury acetate and oleate, usually at concentrations of about 0.1% to achieve stain control on freshly felled green wood. These compounds are still used for this purpose in some areas but they are very toxic and were largely replaced by sodium pentachlorophenate after about 1940. The more volatile compounds such as ethyl mercury acetate possess limited effective life and, where organomercury compounds continue to be used, phenyl mercury acetate or preferably oleate are normally employed. Pyridyl mercury chloride and stearate have also been proposed, and diphenyl mercury and phenyl mercury chloride have been reported as possessing good preservative activity against termites.

Organotin and organolead compounds

Silicon, germanium, tin and lead form Group IV/IVb in the periodic classification. Generally, the stability of organic compounds of these metals decreases with increasing atomic weight from silicon through germanium and tin to lead but in other respects the metals behave in a fairly analogous fashion to give four types of tetravalent compounds; RMX_3, R_2MX_2, R_3MX and R_4M, where M is the metal, R is an alkyl or aryl group and X is the 'anionic radical' which differs from the R groups as it is not attached to the metal by a C–M bond.

Biological activity increases steadily from germanium to lead, being moderately developed in R_2MX_2 and highly developed R_3MX, the silicon compounds and the other two structures exhibiting virtually no activity. The greatest activity is associated with the tin and lead compounds but, as lead is far less expensive than tin, it would appear to be the more economic. In fact, the lead compounds tend to be more reactive, introducing manufacturing difficulties and instability in use so that the tin compounds are generally preferred. In addition, residues from the degradation of organolead compounds are toxic inorganic compounds, although formed in only insignificant amounts from biocidal treatments, whereas the ultimate degradation products from tin compounds are generally innocuous.

Triphenyltin compounds have found use in agriculture and triphenyllead compounds have been used in anti-fouling compositions but, in other biocidal applications, the trialkyltin compounds have been most widely used. Their microbiological activity depends on the total number of carbon atoms in the alkyl chains for both symmetrical and asymmetrical compounds – the greatest fungicidal activity develops with 9–12 carbon atoms, an observation made in about 1952 which prompted proposals for their use as wood preservatives. The mammalian toxicity decreases sharply with increase in total carbon atoms, perhaps being related to the water solubility of the stable hydrolysis products – trimethyl and triethyl compounds hydrolyse to the water-soluble and highly toxic hydroxides, tripropyl compounds form stable oxides or hydroxides, whilst tributyl and longer chain compounds form oil-soluble oxides of lower toxicity, and trioctyl compounds are completely non-toxic. The insecticidal properties decline steadily as the number of carbon atoms increases and reduce sharply when these exceed about 15, while general wood preservative activity decreases conversely to the increasing equivalent weight, and then decreases more sharply in excess of 15 or 18 carbon atoms.

Tributyltin compounds

Tributyltin compounds are therefore preferred as they offer the greatest separation between mammalian toxicity and useful biocidal or preservative action, and tri-*n*-butyltin compounds are the only organometallic compounds of Group IV metals that are extensively used in wood preservation. Although they were proposed as wood preservatives following observations of their exceptional fungicidal and insecticidal activity, they are actually non-toxic when applied to wood at low retentions, yet the wood does not necessarily decay. It appears that the tributyltin group may have a chemical affinity for wood cellulose, causing modification which inhibits decay, although fungal hyphae may penetrate into the treated wood. There is thus a distinct danger that internal decay can occur if the preserved zone is relatively shallow, a danger that is normally avoided by using a much higher retention of tributyltin oxide or by the addition of other non-fixing fungicides. For example, a 0.1% organic-solvent solution of tributyltin oxide can be shown in laboratory experiments to give protection to a completely impregnated pine block but in commercial treatments 1% is more normal, or alternatively 0.5% when other fungicides are present.

Experience since 1959 when tri-*n*-butyltin oxide was first introduced commercially in the United Kingdom suggests that the addition of other fungicides is advantageous, particularly pentachlorophenol, *o*-phenylphenol and borates as these appear to give considerably improved resistance to White rots, whereas the organotin compounds give particular protection against the Brown rots which attack only the cellulose in wood. These mixed formulations are particularly preferred for remedial treatment preservatives whilst the higher concentration of 1.0% tri-*n*-butyltin oxide is now used extensively for the treatment of external joinery (millwork), particularly by the double vacuum process. Formulations sometimes also contain contact insecticides, although tri-*n*-butyltin oxide is distinctly insecticidal and can give excellent protection on its own when applied at adequate retentions.

The House Longhorn beetle is readily controlled in this way at retentions of about 1 kg/m³ whereas twice this concentration or more is required to control the Common Furniture beetle. Numerous proprietory organic-solvent wood preservatives now contain tri-*n*-butyltin compounds as their principal toxicants, and such formulations are extensively applied by pressure impregnation, double vacuum, immersion or spray.

The volatility of tri-*n*-butyltin oxide can be troublesome during hot weather and less volatile compounds have been used such as naphthenates and phosphates, although they all eventually hydrolyse to the oxide. Problems at low temperatures usually indicate contamination with more volatile compounds such as halides through poor manufacture or reaction with additives; some stabilizers that are added to TBTO can cause increased volatility. There is also some evidence of degrade on treated wood to di- and monobutyltin but, whilst these forms have lower microbiological activity, there is no evidence that the long-term preservative action has been affected, probably because it depends on blocking hydroxyl groups on cellulose and losses by volatilization and degrade affect only the excess unreacted compound.

The fungicidal preservative action of tri-*n*-butyltin oxide is enhanced when it is applied in the presence of a swelling solvent, particularly water. Permapruf T, known in the Nordic area as BP Hylosan PT, was the first proprietary pretreatment product to take advantage of this observation and consisted of tri-*n*-butyltin oxide solubilized in water using quaternary ammonium compounds. While other surfactants can be employed, they are generally less reliable and cannot contribute to the preservative activity in the same way. A sapwood retention in pine of 1.2 kg/m³ TBTO in Permapruf T is equivalent to the normal required retention of 9 kg/m³ active oxides in conventional CCA formulations, which is achieved with retentions of 15 kg/m³ Tanalith C or Celcure A; the overall retentions are approximately half these figures. These retentions, based on stake trials, suggest that Permapruf T, which contains 10% TBTO, is about 25% more effective than a British Standard 4072 CCA salt product, such as Tanalith C and Celcure A, which contain about 60% active oxides.

The systematic studies by the author in 1960–70 on the organic compounds of the Group IV/IVb elements silicon, germanium, tin, and lead prompted interest in similar studies on other groups in the periodic classification. In Group III boron has been extensively used, as borates derived from boric acid but not as organic compounds, but aluminium has proved useful in water-repellent and film-forming compounds as described later in this chapter. Phosphorus, sulphur and chlorine from Groups V, VI and VII are used as component elements in many of the modern complex organic fungicides and insecticides, but only Group V gives rise to a range of compounds which have similarity to the organometal compounds of Group IV. Group V/Vb comprises nitrogen, phosphorus, arsenic, antimony and bismuth, but only the great variety of organic compounds of nitrogen have attracted special attention.

Organonitrogen compounds – quaternary ammonium compounds – alkyl ammonium compounds (AAC)

The organic compounds of nitrogen are generally known as amines and do not usually have any special biocidal activity, but the quaternery ammonium compounds, sometimes known as alkyl ammonium compounds (AAC), are distinctly different. The greatest biological activity is associated with compounds with a single anionic group such as chloride or bromide and a cation comprising nitrogen with four organic groups. The simplest cation is ammonium, NH_4. With larger organic groups water solubility is associated usually with a benzyl group and these compounds are favoured for wood preservation as they avoid the need for a co-solvent such as an alcohol. The precise structure of the other three organic groups attached to the nitrogen is relatively unimportant in the sense that greatest biocidal activity develops when the carbons in these groups total about 16. An alkyl–benzyl–dimethyl ammonium compound with an alkyl group with a chain length of about 14 therefore represents optimum fungicidal and bactericidal activity; compounds of this structure are known as benzalkonium compounds and are used as antiseptics. They are very effective as wood preservatives, achieving an eradicant action but also a persistent action through condensing onto hydroxyl groups on wood cellulose. A compound with a slightly different structure, benzalkyl-trimethyl ammonium, in which the benzalkyl structure forms a single group on the nitrogen, has been extensively used in wood preservation, particularly in Europe where it was known as Gloquat C but it has now been withdrawn as there are health problems associated with the preparation of the raw materials from which it was manufactured.

Preservative activity has been reported for a wide range of amines and quaternery ammonium compounds but this activity is associated in many cases with a surfactant effect or degrade to ammonia with no true preservative action. The failure of some of these compounds has led to doubts regarding the reliability of the quaternery ammonium compounds but, if compounds of suitable structure are selected, they have a broad spectrum of activity, although some of the compounds can be degraded by resistant fungi. It is therefore advisable for quaternery ammonium compounds to be used only in association with other fungicides such as borates.

4.6 Carrier systems

The carrier system is as important as the toxicant system in a wood preservative formulation. The advantages and disadvantages of various systems have been discussed theoretically in Chapter 3 and there are also appropriate notes in Section 4.2 of this chapter. However, formulated preservatives need a carrier solvent, the most important choice being between polar solvents such as water and alcohols, which have reactive hydroxyl groups which will cause swelling in wood, and non-polar solvents such as petroleum distillates which will avoid swelling.

Organic solvents – proprietary advantages

Whilst an organic solvent may not contribute directly to the preservative properties it is certainly of considerable importance. Non-polar light petroleum distillates have low viscosities and are able to penetrate rapidly into dry wood so that they are particularly suitable for use in preservative formulations that are designed for superficial application by brush, spray or immersion. However, it is not sufficient to achieve deep penetration by the preservative formulation as the subsequent volatilization of the solvent may cause the toxicants to return to the surface where they will be particularly susceptible to losses by volatilization and leaching. Co-solvents are frequently used, sometimes to assist in the solution of the toxicants during manufacture but often to provide high viscosity residues or tails which will retain the toxicants whilst the carrier solvent volatilizes, ensuring proper distribution. In some cases the use of co-solvents distinctly reduces the apparent toxicity of an active component, perhaps due to deeper distribution or a protective action which also ensures greater persistence and a longer effective life. For example, pentachlorophenol is often used in heavy oil as an alternative for creosote, the heavy oil persisting and protecting the pentachlorophenol from leaching and volatilization. While many products may be based on the same concentration of a particular toxicant, their performance may vary widely and it must not be assumed that apparently similar proprietary products will achieve similar performance.

The penetration of polar organic solvent systems can be improved by the addition of various resins and waxes, particularly by adding stearic or palmitic acids or esters; the latter act as surfactants, condensing onto the hydrophilic wood components and permitting an hydrophobic solution to penetrate. Surfactant systems of this type are extensively used as additives to road tar and bitumen to assist wetting of damp crushed stone.

There have been many attempts to use water more extensively to avoid the high cost of organic solvent systems and to reduce fire and health risks. In suspension systems, liquid or solid toxicants are dispersed directly in the water using only a surfactant; the toxicants do not penetrate and are simply deposited on the wood surface. In emulsion systems the toxicants are generally dispersed as a concentrated solution in solvent emulsified in water. When applied to the surface of wood the emulsion breaks, allowing the solvent phase to penetrate whilst the water evaporates. The fire risk associated with the use of sprayed organic-solvent preservatives is considerably reduced but in spray application the penetration depends directly on the solvent concentration that is present and can never be as good as with organic solvent systems.

Bodied mayonnaise-type emulsions (BMT) – Woodtreat

One problem in remedial treatment preservation is the deep penetration that is required where large wood sections are involved which may be suffering from deep borer infestations or fungal infections, and in the treatment of external joinery which may also be suffering from internal fungal decay, concealed by a paint coating. The bodied mayonnaise-type (BMT) emulsion products are pastes which can be applied to the surface of wood with a trowel or gun, the emulsion breaking in contact with the surface so that the organic-solvent phase containing the toxicants is able to penetrate slowly. Organic solvents of low volatility are used as very deep penetration can be achieved in this way, although many proprietary products contain inadequate concentrations of toxicant – the concentration must depend on the required retention related to the degree of penetration that may be achieved. The best known product of this type is Woodtreat.

Wykamol injectors – borate diffusion treatments – Boracol – Borester 7 – Timbor rods

Wykamol remedial products were usually applied in the past by drilling deep holes which were then injected under pressure using a conical nozzle. This system has been replaced by a moulded plastic injector which is driven into each hole, leaving a projecting nipple to which a pressure gun is fixed. Organic-solvent products are appreciably compressible and the injector is fitted with a non-return valve, thus trapping within the hole a reservoir of preservative which is then able to disperse deeply within the wood. Unlike the BMT systems this technique can be used for the treatment of window frames and other external joinery; after treatment is complete each nipple is removed with a chisel and the hole stopped and painted to conceal the remains of the injector within the wood. Concentrated borate solutions such as Boracol 20 and 40 and Borester 7 can be used in downward holes, slowly diffusing; Timbor rods can also be inserted into holes, diffusing if the wood becomes wet. The borate products are described earlier in this chapter.

Smoke treatments

Insecticidal smokes must also be mentioned as they are claimed to have advantages over the use of formulated organic-solvent preservatives. A smoke can achieve only a finely dispersed solid deposit concentrated largely on upper horizontal surfaces. With contact insecticides such as Lindane or Dieldrin any insect settling on the treated surface will be killed, but the finely dispersed and superficial nature of the deposit ensures only transient protection.

141

Gas treatments – methyl bromide – Dichlorvos – methyl borate

Fumigant gases are attractive because of their very low viscosities compared with liquids and the penetration that they can achieve. Methyl bromide is very effective in eradicating insect infestations such as termites but it is extremely poisonous. A stack of wood or an entire structure such as a house must be sealed for several days to permit the methyl bromide to entirely eradicate any deep-seated borer infestation as there is no residual action, and the gas must then be entirely removed by ventilation. An alternative system for a building is to install porous strips impregnated with a volatile insecticide such as Dichlorvos, although the strips must be replaced annually, shortly before the flight season, as with a smoke treatment, and a more realistic use for this particular insecticide is at low concentrations in multicomponent products as an initial eradicant insecticide. Methyl borate is a much more realistic gaseous preservative. Although it will penetrate deeply into wood it will subsequently hydrolyse to deposit boric acid, but realistic retentions of boric acid can only be achieved if this preservative is applied by impregnation processes to achieve deep penetration, the gaseous phase only assisting with subsequent diffusion.

4.7 Water repellents, stabilizers and decorative systems

Changes in moisture content up to fibre saturation point invariably involve movement, shrinkage with drying and swelling with wetting. Although it is normal to dry wood to a moisture content equivalent to the average atmospheric relative humidity anticipated in use, it is common to encounter movement problems. Faults such as gaps appearing between floor blocks or boards are due to the wood drying after installation, either through inadequate kilning or perhaps re-wetting between kilning and in-stallation. A door or drawer jammed in humid weather may be exceedingly slack under drier conditions. Frames which introduce an end-grain surface in contact with side-grain will inevitably result in cracking of any surface-coating system. In other situations the cross-sectional movement may become apparent as warping through twisted grain effects. The obvious solution to all these problems is to use only wood with low movement but this is not always realistic. The alternative is to impregnate the wood with chemicals which induce stabilization. Unfortunately these processes are also frequently unrealistic because of the difficulty of achieving complete impregnation, a problem that has already been discussed in connection with normal wood preservation.

Paint and varnish

One obvious solution is to enclose the wood within a protective film to stabilize the moisture content. Paint and varnish coatings will act in this way, provided they completely cover the wood and remain completely undamaged. Unfortunately, whilst these coatings give good protection against rainfall, they are unable to prevent moisture content changes resulting from slow seasonal fluctuations in atmospheric relative humidity. As a result the painted wood will shrink or swell with changes in relative humidity, causing the surface coating to fracture wherever a joint involves stable side-grain in contact with unstable end-grain. Rain is absorbed by capillarity into the crack, yet the remaining paint coating restricts evaporation, so that the moisture content steadily increases until fungal decay is sure to occur if the wood is non-durable. It is frequently suggested that preservation provides a simple solution to this problem, but this ignores the fact that water also damages the paint coating. It is explained in Chapter 2 that wood is an hygroscopic material, covered with hydroxyl groups which have a strong affinity with water so that penetrating water will tend to coat the wood elements, displacing paint and varnish coatings. This failure is known as preferential wetting and is responsible for blistering and peeling in paintwork and the loss of transparency in varnishes.

Water repellent preservatives

The best solution to both the decay and preferential wetting is treatment of external joinery (millwork) and cladding before painting, with a formulation that is both a preservative and a water repellant. A water-repellent treatment coats the pores of a structural material, reversing the angle of contact so that capillary absorption of water is prevented; water repellency is associated with water globulation on the surface but absence of globulation on a weathered surface does not necessarily mean that a treatment is no longer effective as the pores may still be water repellent.

Waxes and resins

Various waxes, particularly paraffin waxes, are the most commonly used water-repellent components in wood preservative formulations, although a treatment based on a wax is generally as susceptible to preferential wetting failure as the surface coating that it is designed to protect. Treatments of this type are reliable only if they penetrate deeply and are applied at sufficient retentions to ensure that the wood elements are entirely inaccessible to even changes in the relative humidity of the atmosphere. High-wax retentions cannot be used if wood is to be finished with a paint or varnish coating as the adhesion is seriously affected – the coating cannot adhere to the wax deposit and is unable to penetrate if the wax retention is too high, but in addition, the wax may migrate into the coating solvents, affecting both solvent loss and the ability to absorb the oxygen required for drying so that the coating may remain tacky. The wax will continue to migrate through subsequent coatings, affecting inter-coat adhesion, perhaps even causing cissing, the situation when a coat is unable to wet a surface and tends to concentrate in globules, leaving other areas uncoated. Yet another problem with migrating wax is the tendency to prevent the development of gloss in the final top coat. For these various reasons waxes are generally used at low retentions and the desired pore-sealing action is achieved by the addition of resins.

Resin selection is critical in terms of water and water vapour resistance as well as paintability. The aliphatic and aromatic hydrocarbon resins are inexpensive and efficient but they do not dry; they solidify only by loss of solvent and may be re-dissolved by coating solvents, perhaps interfering with the drying and durability of the coating system. Natural drying oils such as boiled linseed oil can also be used but paintability problems may arise through slow drying. The use of suitable modern alkyd resins can avoid this difficulty but they are also expensive. The most realistic systems therefore tend to be based on mixtures of waxes, hydrocarbon resins and alkyd resins to avoid these problems, and there are therefore distinct differences between proprietary products.

It is particularly important to appreciate that unsaturated or drying resins are likely to significantly reduce the activity of some cationic preservatives such as zinc, and particularly tributyltin oxide. In addition, alkyd resins will solidify only in the presence of driers or catalysts such as metal naphthenates, and these catalysts may be inactivated by tributyltin oxide which is a base and will thus absorb their acids. Such problems can be avoided by using other toxicants, yet tributyltin oxide is particularly suitable as it tends to improve the resistance of the formulation to preferential wetting, and it is therefore better to use tributyltin compounds other than the oxide such as the naphthenate or o-phenylphenate which do not suffer from these disadvantages.

Silanes (silicones)

There have been several attempts to develop more suitable water repellent components in view of the difficulties associated with the use of waxes. Tributyltin oxide orientates onto the wood fibres, giving a water-repellent surface through the presence of the hydrophobic butyl groups. However, this compound is expensive, and toxic at high retentions, so that use as a water repellant is unrealistic, but other Group IV organometal compounds can be used. The organosilicon compounds, the silanes or silicones, are the best known water repellents in this group but the very stable silicone oils tend to possess many of the disadvantages associated with heavy organic oils and waxes. The only silicones suitable are those which have a high degree of functionality so that they are able to attach themselves to the wood components in the same way as tributyltin oxide, thus giving good resistance to preferential wetting failure. They have not been extensively employed, probably through disappointing results following the use of unsuitable silicone oils and resins.

Organoaluminium compounds – Manalox

Organic compounds of aluminium, titanium and zirconium can also be used but the water-repellent groups in typical available commercial products are usually long-chain fatty acids such as stearate which give a waxy treatment and are more susceptible to oxidation when applied at low retentions than the short-chain alkyl groups on typical silicone resins. However, aluminium compounds can incorporate unsaturated chains and, when used for preservative, water-repellent or priming treatments, they can provide excellent adhesive bonding between the wood elements and alkyd systems, giving resistance to preferential wetting. Even toxic groups such as pentachlorophenate can be incorporated, thus avoiding the need for special co-solvent or anti-blooming systems. These principles are most highly developed in various Manalox products which can be described as polyoxoaluminium compounds. These advantages of organoaluminium compounds are not apparent in the normal aluminium stearate, which performs only in the same way as a wax.

Stabilizers

If tributyltin oxide is applied at retentions in excess of the toxic limits required to protect wood against fungal decay, the treatment eventually saturates all the free hydroxyl groups on the cellulose chains which are responsible for hygroscopic movement and the wood becomes completely stabilized. Such treatments are uneconomic but there are other possible systems for chemically reacting these troublesome hydroxyl groups. Formaldehyde treatment in the presence of an acid catalyst will cross-link hydroxyl groups on adjacent chains, reducing the dimensions of the wood in the process but also reducing the movement to less that 10% of normal. Acetylation involves the treatment of wood with acetic anhydride in the presence of a strong acid catalyst, a process that considerably reduces the hygroscopicity of wood and also increases its resistance to fungal attack. However, all these chemical modification treatments suffer from the severe disadvantage that they are effective only if the wood is completely impregnated and they can therefore be used realistically only on permeable species; acetylation is being used in this way to an increasing extent on radiata pine.

Bulking – Impreg – PEG – Carbowax – MoDo

In bulking, the wood is impregnated with a very high retention of material which will physically restrain movement. Several resin systems have been employed in this way such as the phenolic resin in Impreg and a styrene/polyester co-polymer system used for the impregnation of floor blocks, in Finland. These systems rely on physical restraint and are reliable if deep penetration is achieved, although complete penetration is not essential. The polyethylene glycol waxes are also bulking treatments but they are applied in water and retain the wood in the expanded wet state. Treatments of this type such as PEG, Carbowax and MoDo are generally applied by prolonged diffusion. The compounds with low molecular weight of 200–600 are readily soluble in water and diffuse reasonably quickly but 1000 is less soluble and gives slower diffusion, although it is less hygroscopic so that after drying the wood is not so tacky as with the lower molecular weight treatments. These systems are used particularly for the stabilization of archaeological specimens, the largest to be treated so far being the warships *Wasa* in Stockholm and *Mary Rose* in Portsmouth which were spray-treated with a mixture of polyethylene glycol and borate whilst the atmospheric relative humidity was maintained at a high level to prevent drying. Although this system has been used successfully for the stabilization of gun stocks, the relationship between molecular weight, treatment time and hygroscopicity is a distinct disadvantage. In one system for the treatment of floor blocks the low molecular weight compounds are employed, followed by complete drying, and the introduction of isocyanate vapour which reacts with the glycol to form a polyurethane resin, avoiding all the disadvantages and giving a treatment which is stable and resistant to heavy floor wear. There are many other polymer systems that have been or could be used but they are generally unrealistic, combining the need for high retentions with expensive chemical compounds.

Decorative preservatives – Madison formula

While many water-repellent preservatives are designed specifically for use as pretreatments prior to painting or varnishing, perhaps in place of conventional priming treatments, other systems are designed as complete maintenance treatments, frequently serving a decorative as well as a protective function; these decorative preservatives are particularly popular in the Nordic countries. The two types of water-repellent preservative are not necessarily similar; the first type must be compatible with subsequent paint or varnish coatings whilst the second type must clearly have good resistance to weathering. The Madison formula, developed in the United States as a maintenance treatment for western red cedar cladding, is perhaps the best known. It consists of paraffin wax, pigments and boiled linseed oil binder with pentachlorophenol as the preservative and zinc stearate to give water repellency, colour retention and freedom from stain. It has now been largely replaced by various improved proprietary products, those containing trihalomethylthio-compounds being much more efficient in controlling stain, as explained in Section 4.9.

Royal process

Weather resistance is poor with systems that are simple deposits of hydrophobic components such as waxes which are susceptible to preferential wetting, but can be improved by using a binder as in the Madison formula or by fixation to the wood as with silicone resins, although deep penetration will improve the performance of most systems. In the Royal process developed by Häger for the treatment of external joinery (millwork) a water-borne preservative treatment is followed by deep treatment with a drying oil. This is a very effective process but involves a complex multi-stage treatment and the need for a multiple oil-storage system to provide finishes in different colours. Whilst the Royal process gives an exceptionally durable decorative finish it is also very expensive. The main problem is that treatment is carried out in two stages, the first introducing large quantities of water which must be removed before the second oil impregnation stage can be satisfactorily achieved. There is no reason why similar reliability could not be achieved by single impregnation with an organic system, designed to achieve both the preservative and the decorative functions, but commercial companies are reluctant to invest in systems that, because of the need for different colours, involve multiple storage tanks and a danger of contamination in the impregnation cylinder.

4.8 Fire retardants

Modern water-borne fire-retardant formulations originated in 1821 when Gay-Lussac reported that ammonium phosphate, ammonium phosphate with ammonium chloride, and ammonium chloride with borax (sodium tetraborate) were excellent fire retardants when applied at adequate retentions to cellulosic fibres. These substances were not so effective when used on wood but this was due to difficulty in achieving adequate retentions from superficial treatments. They were far more effective when applied by pressure impregnation to give higher retentions but, as with the salt preservatives, corrosion problems were encountered and the use of ammonium sulphate was discontinued for this reason shortly after it was introduced in 1880.

Oxylene – Minolith – Celcure F – Pyrolith – Fyre Prufe – Minalith – Pyresote

The Oxylene process was introduced in 1905, followed by Minolith in about 1915 which consisted of Triolith wood preservative with the addition of a large concentration of rock salt, to give a combined preservative and fire retardant for use in mines. In about 1930 Celcure F was developed, in which the acetic acid in normal Celcure was replaced by boric acid which, with added phosphates and zinc chloride, performed well as a flame retardant. Various competitive systems followed such as Pyrolith, Fyre Prufe, Minalith and Pyresote, all based on similar mixtures of soluble salts, usually added to established preservative formulations.

Fire-retardant salt components are leachable but also hygroscopic so that normal coating systems cannot be used on treated wood to give protection against leaching. The compositions of the products have varied with the availability of individual chemical compounds. The most popular components are ammonium phosphates, ammonium sulphate, zinc chloride, boric acid and borates. A fire retardant must suppress both flaming and after-glow but only a few compounds can achieve this when used alone.

Non-Com

Generally, formulations containing zinc chloride such as Pyresote, which also contains ammonium sulphate, boric acid and sodium dichromate, and might be described as a fire-retardant version of chromated zinc chloride, are declining in use, whereas the less sophisticated mixtures of ammonium phosphates, ammonium sulphates and borates are becoming more popular. More expensive systems such as Non-Com, which polymerizes within the wood, have been introduced to avoid the problems associated with soluble and hygroscopic salts. This improvement has certainly increased the scope of fire-retardant treatments but it is clear that there is far more interest in North America than in Europe.

Halogenated compounds

An alternative method for achieving leach resistance is to use only water-insoluble organic compounds. Generally, fire retardants can be prepared by using high loadings of halogenated compounds such as chloronaphthalenes and chlorinated paraffins, although their effectiveness as fire retardants is greatly increased if they incorporate catalysts, particularly antimony or zinc compounds. Brominated compounds are also used. These are the only formulations that may be suitable for remedial use in building structures, but it is difficult to find solvents that do not themselves introduce a fire hazard during the application process. The very high retentions that are necessary unfortunately make these treatments very expensive and they have not been used to any significant extent. In recent years more sophisticated organic polymer treatments have been developed, primarily for the use on textiles – whilst these are efficient on wood, they again suffer from relatively high cost and the need for high retentions. It is therefore usually more realistic to use intumescent systems for in-situ treatments but, as these are essentially coatings rather than preservatives, their composition is not considered in detail here; their mode of action has already been discussed in Chapter 3.

4.9 Stain control

Organomercury compounds – sodium pentachlorophenate

The normal wood preservative systems give comparatively poor control over the sapstain fungi and superficial moulds that are principally responsible for stain in freshly felled green wood, and under coating systems in service. It was eatablished in about 1935 that organomercury compounds were extremely effective in controlling these mixed fungal infections and chlorophenols were introduced about five years later. The organomercury compounds are very toxic and, despite various attempts to reduce mammalian toxicity, by replacing ethyl mercury by phenyl mercury compounds and acetates by oleates, their use is now forbidden or officially discouraged in most countries. Despite the irritant nature and moderately high toxicity of sodium pentachlorophenate, it remains in use in most countries as efficient and economic alternatives are not readily available. However, this compound is not permitted in some countries such as Sweden, where there are fears of health risks associated with dioxin impurities, as described in the earlier section on organic compounds.

Fluorides and bifluorides

Fluorides and particularly the so-called bifluorides such as ammonium hydrogen difluoride have been extensively used for sapstain control but they are not very reliable. Ammonium hydrogen difluoride gives good control of stain fungi but actually stimulates growth of some surface moulds such as *Trichoderma viride* which can cause severe problems unless very high fluoride concentrations are used.

Pentabor

In the absence of any obvious simple alternatives co-formulations with other compounds were developed, originally as a means to reduce the toxicity of the organomercury compounds and the irritancy associated with sodium pentachlorophenate. Some co-formulations, particularly those based on combinations of sodium pentachlorophenate and borax, have been widely used throughout the world and offer treatments of lower toxicity than with sodium pentachlorophenate alone. Co-formulations suffer from the obvious disadvantage that several separate components must be measured, perhaps in small quantities, when topping up treatment tanks. The most popular co-formulation consists of 1 part sodium pentachlorophenate with 3 parts borax (sodium tetraborate decahydrate). The cost of manufacture has tended to discourage the use of ready mixed compositions but Pentabor, a co-formulation of this type developed in England, has half the water of crystallization removed, to reduce transport costs. Bromophenols have also been used, particularly tribromophenol, both alone and in combination with borates, but they have no significant advantages over the less expensive and more effective chlorophenols, except that their toxic dangers are less well known and their use is therefore less restricted.

Borates

Chloro- and bromophenols are generally used as sodium or potassium phenates, the alkalinity or high pH greatly enhancing their stain control activity. The addition of inactive sodium carbonate prolongs control, apparently by maintaining this high pH. Borax was originally added to sodium pentachlorophenate to broaden the spectrum of activity but the ability of borax to maintain the high pH is probably equally important. Borate compounds cannot be used alone in stain control treatments because they are not effective against some surface moulds, but they are very effective against sapstain itself and are now used as the main toxicant in many formulations, additional fungicidal components effectively controlling only the surface-mould problem.

Benomyl (Benlate) – Captofol (Difolatan) – quaternary ammonium compounds – zinc borate – Polyphase (IPBC) – Chlorothalonil (Tuffbrite) – isothiazolones (ITA) – thiazoles (MBT, TCMTB) – triazoles (Madurox)

Suspensions of Benomyl (Benlate) and Captafol (Difolatan) were introduced as alternatives to chlorophenols but they are not generally very effective; they leave only a superficial deposit which, whilst it may control surface stain and mould, has no influence over deep stain within the wood. These fungicides are therefore most reliable when formulated with borates which will control deep sapstain. The most effective stain control treatments are therefore mixtures of borates which will control deep sapstain with other fungicides such as Benomyl, Captafol or even quaternary ammonium compounds which will control surface mould growth. Completely inorganic systems such as zinc borate formulations are not currently used but are probably the most promising sapstain control treatments for the future. This development has been largely ignored, probably because it is well known that cations such as copper and zinc are not effective alone against sapstain, and chemical manufacturers have therefore concentrated on research on increasingly complex organic compounds, which are generally expensive and lack permenance when widely dispersed on a wood surface exposed to strong sunlight, as a stain control treatment. Some of these organic compounds have proved effective and marketable for stain control. They are all described in more

detail in the earlier section on organic compounds. Carbamates such as IPBC (Polyphase), chlorothalonil (Tuffbrite), various isothiazolone (ITA) compounds, thiazoles such as MBT and TCMTB, and triazoles such as cebuconazole, propiconazole and azaconazole (Madurox) have performed well in trials, alone or in various formulations.

Stain in service (under coatings) – trihalomethylthio- compounds – Captan – Folpet (Fungitrol 11) – Fluorfolpet – Dichlofluanid (Preventol A3, A4)

It will be appreciated from the comments in Chapter 3 that the control of stain under paint and varnish on exterior joinery (millwork) is essential to achieve a reasonable life for a decorative coating system. One of the best ways to apply a stain control treatment is as a normal wood preservative, perhaps in a water-repellent or priming formulation, as a pre-treatment prior to paint or varnish. Normal organic-solvent wood preservatives possess poor activity against stain fungi and mould, even if they contain pentachlorophenol at 5% or tributyltin oxide at 2%, although both these toxicants are also used for mould control in emulsion paints. Copper 8-hydroxyquinolinolate and Thiram are more efficient, whilst diphenyhl mercury dodecenyl succinate (Nuodex 321 Extra) gives excellent results initially, but lacks persistence. Only the trihalomethylthio- compounds have proved consistently reliable. Captan is probably the least efficient of this group of compounds but this may be only a solubility factor as Folpet (Fungitrol 11) with the same trichloromethylthio- radical is very active. The dichlorofluoro-compounds Fluorfolpet (Preventol A3) and Dichlofluanid (Preventol A4) are also very effective and are preferred by many manufacturers as they are more readily soluble in most organic solvents, whereas Captan and Folpet are virtually insoluble and can usually be applied only in relatively high-viscosity systems, such as pigmented coatings, or formulations with a relatively high resin content. However, none of these treatments give permanent protection as they are lost by volatilization and oxidative degradation, in common with almost all organic compounds; only inorganic systems such as zinc borate formulations are able to provide permanent protection.

Stain in service is perhaps most apparent when simple clear or pigmented preservative systems are applied to external cladding and joiner (millwork). The pigmented and water-repellent Madison formula was introduced in the United States some years ago as a maintenance treatment for western red cedar cladding, relying on pentachlorophenol and zinc toxicants. Performance has been unpredictable – the formula contains boiled linseed oil which is readily attacked by some of the stain fungi on wood, particularly *Aureobasidium pullulans*, and there have been many attempts to develop improved proprietory products. In some cases the risk of stain and mould development has been reduced by the use of alternative binders, such as the Manalox compounds described as water repellants earlier in this chapter, but a great variety of toxicants has also been used. Some of these have been introduced without proper evaluation whilst others, such as the organomercury compounds, have shown good initial activity but poor life. Some proprietory treatments have actually increased the stain risk! The most successful products generally contain one of the trihalomethylthio- compounds.

4.10 Remedial treatments

Remedial treatment preservative formulations have been mentioned up to now only in passing as they actually represent a distinct and rather specialized development route. They will be described only briefly; a more detailed account is available in the book *Remedial Treatment of Buildings* by the present author.

Xylamon – Rentokil – Wykamol (Anabol)

Few products were produced specifically for this purpose before about 1920. Xylamon was introduced in Germany in about 1923 as an eradicant for House Longhorn beetle using mono- and dichloronaphthalenes with their characteristic pungent odour. This development was followed in the United Kingdom by the introduction of Rentokil based on o-dichlorobenzene, and in 1934 by Anabol, later called Wykamol, which consisted principally of chloronaphtalene wax with a small amount of o-dichlorobenzene. From that point Richardson & Starling Limited, the manufacturers of Wykamol, became leaders in the development of new remedial treatment preservatives. In 1939 the o-dichlorobenzene in Wykamol was replaced by Rotenone, a natural insecticide extracted from Derris root. DDT was considered as a replacement when it was introduced during World War II but it was not very effective. In 1945 the Rotenone was replaced by Lindane which was found to be far more effective, particularly in combination with chloronaphthalene wax which tended to protect it from volatilization, increasing its effective life without significantly reducing its activity.

Cuprinol – Reskol

The Cuprinol copper and zinc naphthenate products had been widely used previously for the eradication of fungal infections in building timber, although failures occasionally occurred as it is difficult to achieve adequate retentions by superficial brush or spray application. In 1936 Richardson & Starling introduced Reskol which was designed specifically as an eradicant fungicidal wood preservative. It consisted originally of barium naphthenate and p-dichlorobenzene in light creosote. In about 1955 the formulation was changed to 5% pentachlorophenyl laurate in a light petroleum distillate but this proved rather unsatisfactory, although it is still used by other manufacturers. It was therefore replaced in 1957 by 5% o-phenylphenol; many other manufacturers have used 5% pentachlorophenol but it is very unpleasant to apply by spray during remedial treatment.

Insect infestations are often encouraged by or dependent on fungal decay and a combined insecticide and fungicide treatment was therefore introduced by adding pentachlorophenyl laurate and later o-phenylphenol to Wykamol, resulting in Wykamol PCP and Wykamol Plus. This combined product eventually replaced Wykamol and Reskol. In 1961 tri-n-butyltin oxide was added, originally as a persistent fungicide, replacing a proportion of the o-phenylphenol. Wykamol Plus has continued to contain two complementary fungicides and has thus largely avoided the problems encountered by products which rely upon tri-n-butyltin oxide alone. o-phenylphenol was derived from the manufacture of phenol by the Dow process and became scarce and expensive when this process was replaced by the cumene method. This stimulated yet another new development with the replacement of the o-phenylphenol by a borate ester.

In 1967 Wykemulsion was introduced, incorporating the same toxicants as Wykamol Plus but with a reduced amount of solvent emulsified in water. When sprayed on the surface of wood the emulsion breaks, allowing the solvent phase to penetrate whilst the water evaporates. The fire risk associated with the use of sprayed organic-solvent preservatives is considerably reduced. This type of emulsion must not be confused with products containing only a small amount of solvent as these are suitable only for pressure impregnation; in spray application the penetration depends directly on the solvent concentration that is present.

Injection and diffusion treatments – bodied mayonnaise-type emulsions (BMT) – Woodtreat

One problem in remedial treatment preservation is the deep penetration that is required where large wood sections are involved which may be suffering from deep borer infestations or fungal infections, and in the treatment of external joinery which may also be suffering from internal fungal decay concealed by a paint coating. The bodied mayonnaise-type (BMT) emulsion products are pastes which can be applied to the surface of wood with a trowel or gun, the emulsion breaking in contact with the surface so that the organic-solvent phase containing the toxicants is able to penetrate slowly. Organic solvents of low volatility are used as very deep penetration can be achieved in this way, although many proprietory products contain inadequate concentrations of toxicant; the concentration must depend on the required retention related to the degree of penetration that may be achieved. The best known product of this type is Woodtreat.

Wykamol injectors – borate diffusion treatments – Boracol – Borester 7 – Timbor rods

Wykamol remedial products were usually applied in the past by drilling deep holes which were then injected under pressure using a conical nozzle. This system has been replaced by a moulded plastic injector which is driven into each hole, leaving a projecting nipple to which a pressure gun is fixed. Organic-solvent products are appreciably compressible and the injector is fitted with a non-return valve, thus trapping within the hole a reservoir of preservative which is then able to disperse deeply within the wood. Unlike the BMT systems this technique can be used for the treatment of window frames and other external joinery; after treatment is complete each nipple is removed with a chisel and the hole stopped and painted to conceal the remains of the injector within the wood. Concentrated borate solutions such as Boracol 20 and 40 and Borester 7 can be used in downward holes, slowly diffusing; Timbor rods can also be inserted into holes, diffusing if the wood becomes wet. The borate products are described earlier in this chapter.

Smoke and gas treatments

Insecticidal smokes and gas treatments are used in remedial treatment in some circumstances but smokes give superficial deposits with only transitory protection and gases area generally eradicants giving no protecting. They are de-scribed in more detail in the earlier section on carrier systems.

This account of remedial treatments has been limited to building treatments. Other remedial treatments, of poles at the ground line using bandages and injection treatments into poles and railway sleepers (ties), are described elsewhere in this chapter and in Chapter 3, together with descriptions of termite remedial treatments using soil poisoning.

Practical preservation

5.1 General principles

Deterioration risk

Before the most suitable preservation system can be selected the deterioration risk must be clearly defined. With normal structural wood it is possible to define situations where deterioration will certainly occur, and in these severe hazard conditions the use of naturally durable or adequately preserved wood is essential. Perhaps the most important severe hazard situation is ground contact and transmission poles, fence posts, railway sleepers (ties) and construction piles must be properly protected. In most marine situations protection is required against marine borers, and building structures require protection against the House Longhorn beetle and against Dry Wood termites in areas where there is a danger of attack by these insects. In all these situations the use of naturally durable wood is unrealistic as it is both scarce and expensive, and in practice preservative treatment is essential and often required under local regulations.

A moderate hazard exists if deterioration is possible rather than probable. In normal building construction in temperate climates there is the danger that sapwood may become infested by wood-borers and fungal infection may occur if wood becomes wet, perhaps through poor maintenance or condensation. In such situations treatment is desirable rather than essential, although there is one situation in which there may be a stronger case for treatment. External joinery (millwork) such as door and window frames is usually protected by paint or varnish coatings but cracks can develop through movement at joints, permitting the penetration of water and introducing a decay risk. While it is theoretically possible to avoid this danger, by careful maintenance or by the use of wood of low movement, it is clearly desirable to reduce the decay danger by the use of naturally durable or adequately preserved wood, and this is now mandatory in several countries.

Where the deterioration danger is only slight there can be no justification for treatment. The most obvious example is the structural woodwork in a normal dry building. Generally, leaks will become readily apparent before fungal decay can develop and, if the sapwood content of the woodwork is limited, Common Furniture beetle infestation, the normal borer hazard in temperate climates, will be structurally insignificant. In these circumstances treatment cannot be justified; if decay occurs it will be a clear indication of defective design, construction or maintenance. Another slight hazard involves furniture constructed from wood species that are susceptible to attack by Common Furniture beetle or Powder Post beetle; the latter represents a risk only within two or three years after felling.

Stain control

Freshly felled green wood is also subject to a slight hazard, both in the forest and after conversion at the mill. Whilst the moisture content remains high there is a danger that stain fungi can develop and if logs are likely to remain in the forest or in storage for a significant period before conversion the cut ends should be sprayed immediately after felling. Freshly sawn wood must also be treated, although it is frequently claimed that stain can be avoided by rapid drying. In fact, even in kiln-drying there is a danger that stain will develop before the moisture content is significantly reduced and, if kiln-dried wood is wrapped in an attempt to retain the low moisture content, there is a danger that condensation will occur beneath the wrapping and a treatment is still necessary if freedom from stain is to be assured. In practice it is normally more realistic to treat all sawn wood to prevent stain, permitting normal sawn wood for use as carcassing (framing) and other general purposes to dry naturally during storage and transport. Wrapped kiln-dried wood, preferred for joinery (millwork) manufacture, must be treated before and after kilning if complete freedom from staining is required.

Organomercury compounds were used in the past for stain control but the most widely used treatment today in temperate areas is 2% sodium pentachlorophenate solution. In some countries such as Finland only 1% is often used as it is argued that the lower cost and improved safety justifies increased risk of stain development. In Mediterranean-type climates the concentration must be increased to about 3% on softwoods, and in tropical areas about 5% is required on hardwoods. Sodium pentachlorophenate can be replaced by various mixtures incorporating this compound, as described in the previous chapter, but the most efficient consists of 1 part sodium pentachlorophenate with 3 parts sodium tetraborate decahydrate (borax), a mixture that has been in use since World War II and is more effective, more economic and safer than sodium pentachlorophenate alone. Pentabor S is a mixture of this type, which is concentrated by a reduction in the water of crystallization; 1.3% Pentabor S is approximately equivalent in performance to 2% sodium pentachlorophenate. If these mixtures are used on hardwoods in the tropics it is more effective to change the proportions rather than simply increase the use concentrations. Thus Pentabor SA is based on a mixture of 1 part sodium pentachlorophenate with 2 parts borax. When sodium pentachlorophenate is applied to wood the sodium ions are quickly neutralized by the natural acidity, resulting in a relatively superficial precipitate and poor penetration. The mixtures with borax delay this precipitation and thus achieve better penetration and improved control over internal staining which can develop under a superficially treated zone.

Limited penetration is also the main criticism of Benomyl (Benlate) and Captafol (Difolatan) suspensions which have been used in recent years as replacements for sodium pentachlorophenate, although they are effective when mixed with borates which control deep stain. Bifluoride mixtures have also been used such as Improsol and BP Mykocid BS. These solutions and the hydrogen fluoride that they liberate can penetrate very deeply and they give excellent control over internal stain, although they are much less persistent than sodium pentachlorophenate and must also be used at comparatively high retentions to control mould growth so that they tend to be rather expensive.

Pinhole borer control

Pinhole borers will rapidly attack logs while the bark is still adhering. They introduce stain fungi to their galleries, causing both staining and boring damage. Whilst this damage can be prevented by the rapid removal of the bark this encourages stain development so that a stain control treatment is then necessary (Fig. 5.1). It

FIGURE 5.1 Pinhole borer prevention on tropical hardwood logs by spraying with Protostan insecticide after removal of bark. (Stanhope Chemicals Limited)

is often considered more realistic to spray the freshly felled logs, particularly in tropical forests, with a stain control treatment incorporating an insecticide, perhaps as an emulsion or suspension. Lindane has been used for this purpose for many years but it is now being replaced by synthetic pyrethroids such as Permethrin.

Powder Post beetle control

The sapwood of large-pored hardwoods is susceptible to attack by *Lyctus* Powder Post beetle and it is advisable to treat wood with an insecticide (Fig. 5.2) where this is a local danger, even if wood is to be kiln-dried; indeed this insect depends on starch within the wood which tends to be lost during air seasoning, but retained during kiln-drying. Whilst stain control treatment may not be considered necessary when hardwoods are being rapidly kiln-dried an insecticidal treatment is often essential. Borates are particularly suitable for this purpose and are widely used in Australia and New Zealand,

155

FIGURE 5.2 Spraying susceptible tropical hardwood with Lyxastan insecticide to prevent Powder Post beetle attack. Immersion treatment is also used. (Stanhope Chemicals Limited)

and one advantage of the pentachlorophenate and borate mixtures such as Pentabor is the effective control of these Powder Post beetles that can be achieved at the same time as stain control with air-seasoned wood; it is not necessary to add an additional insecticide. Susceptible hardwoods include the temperate oak, elm and walnut as well as the tropical light-coloured hardwoods such as obeche and ramin.

Preservation requirements

True preservation treatments generally involve impregnation using vacuum and pressure sys-

tems. Throughout the world the most important risk situation involves the use of wood in ground contact as there is then a severe danger of fungal decay, principally by Basidiomycetes. These fungi are generally unable to develop when wood is immersed in water but there is then a danger of Soft rot and borer attack in marine situations. Above ground level there is still a danger of decay where rainwater may remain trapped in joints or splits, and this danger can be considerably enhanced where a split occurs through a relatively impermeable coating such as paint or varnish which will tend to restrict dispersion of the water by evaporation. In most temperate areas the sapwood of many hardwoods and softwoods is susceptible to Common Furniture beetle infestation but a much greater structural risk occurs in temperate and tropical areas where there is a danger of infestation by the House Longhorn beetle and Dry Wood termites. Many termites are, however, unable to fly and are more readily controlled by soil poisoning around the structure or the installation of termite shields rather than by treatment of the susceptible wood.

Need for permanence

In all preservation processes there is a need for permanence. Thus all preservatives must have adequate resistance to volatilization and oxidation, depending on the particular conditions of use, and also resistance to leaching if they are likely to be exposed to a high moisture content in exposed or ground contact conditions or in marine environments. While it may appear adequate to enclose wood in a superficial envelope of protective treatment it must be appreciated that deeper penetration will significantly improve the life of treatments which are slightly volatile or water soluble, and natural splits or accidental damage may penetrate through a superficial treatment, perhaps permitting internal decay to develop.

Preservative selection – wood selection

Table A.1 in Appendix A is a schedule of some of the most widely used preservative systems with typical recommended retentions on European redwood or Scots pine. These retentions also apply on most other softwoods, although the actual preservation process may vary depending on the properties of an individual wood species, as shown in Table A.2. For example, Corsican pine *Pinus nigra*, South African pine *Pinus patula*, Shortleaf or Southern pine *Pinus echinata* all possess non-durable heartwood and sapwood but both are also permeable and these species can therefore be readily preserved, al-

though it will be appreciated that the sapwood retentions shown in Table A.1 must be achieved throughout in such woods. European redwood or Scots pine, *Pinus sylvestris*, possesses rather similar properties when fast grown, although in slow-grown wood the heartwood is much less permeable and almost completely resistant to penetration, but also possesses moderate natural durability. It is therefore generally considered that adequate protection can be achieved by treating the permeable sapwood alone. In Douglas fir, *Pseudotsuga menziesii*, the heartwood has good natural durability but the non-durable sapwood is also resistant to impregnation. Adequate penetration can be achieved if the sapwood is incised to give access to routes for tangential penetration.

Unfortunately the same system is less effective on European spruce *Picea abies*; although incising will enable the sapwood to be treated the heartwood is non-durable and there is thus a danger that the outer sapwood will split through shrinkage, following periodic wetting and drying, exposing unprotected and non-durable heartwood. The development of splits is not very likely with spruce as it possesses low movement as shown in Table A.2 and this wood is therefore more suitable for superficial treatment than woods with medium or large movements. If deep penetration into spruce can be achieved with a non-fixed preservative leaching is unlikely following initial drying or seasoning in view of the impermeability of the wood and its natural resistance to re-wetting. This advantage of impermeable woods is often ignored, particularly when considering the reliability of unfixed treatments such as borates, which can be applied readily to green spruce by diffusion to give complete impregnation.

Round poles typically involve relatively impermeable heartwood, perhaps with some natural durability, surrounded by relatively permeable sapwood which enables a preserved zone to be achieved. In sawn wood the surface possesses a random distribution of sapwood and heartwood and, whilst the heartwood may be moderately durable as in slow-grown European redwood, it cannot be readily treated and will deteriorate when exposed to a continuous decay risk such as in ground contact. There is a shortage of redwood of suitable size for transmission poles and a proposal that poles should be constructed by laminating smaller pieces of wood is unrealistic as it exposes moderately durable heartwood which cannot be treated, and which gives inadequate performance when it lacks the protection of surrounding treated sapwood. It is perhaps strange to consider that a more realistic technique might be the use of spruce, incised after lamination, as the low movement would avoid the danger of cracks penetrating through the treated zone.

Full- and empty-cell impregnation

Full-cell pressure impregnation is normally employed when it is required to achieve the highest possible loading of preservative within wood. This system is therefore used to achieve very high retentions of creosote, such as for use in marine conditions, and also for the application of rapid fixing multicomponent water-borne preservatives such as the copper–chromium–arsenic (CCA) types. A full-cell water-borne treatment results in a very high moisture content which considerably increases the weight of wood as well as introducing handling and working problems, but it is normally considered that empty-cell processes cannot be used as fixation will occur within the wood and cause depletion of the active components in the recovered preservative solution. The concentration can be corrected if it is consistent but difficulties also arise through the nature of the fixation process; the recovered solution contains reducing sugars from the cell contents which may cause precipitation within the storage tanks. These problems do not occur with preservatives which fix by loss of a volatile component rather than by reaction with the wood elements; the ammoniacal preservatives and the zinc and copper borate systems which fix by acetic acid loss can be reliably applied by normal empty-cell techniques.

Empty-cell processes are used mainly for creosote treatments for normal ground contact situations, particularly for transmission poles; they achieve preservative retentions that are adequate without wasting excessive preservative. They can also achieve relatively clean treatments, although considerable difficulties are often encountered when using the Rüping cycle which involves an initial air compression stage and, if the pressure of this compressed air is not completely relieved during the final recovery stages, there is a danger that bleeding will continue for a protracted period following treatment. For this reason a Lowry empty-cell cycle is preferred, although it should also be noted that it achieves better distribution of the preservative as well as giving freedom from bleeding. The advantages and disadvantages of these various processes are discussed in Chapter 3.

Double vacuum impregnation – immersion treatment

Where very permeable woods are involved, such as South African pine, or where only limited penetration is necessary as in the treatment of joinery (millwork), low impregnation pressures can be used, even double vacuum treatments which utilize only a vacuum and atmospheric pressure to achieve penetration. It has already been explained in Chapter 3 that, when relatively impermeable woods are involved, pressure increase is not very effective in achieving increased penetration and retention, and to extend the treatment time is far more effective. This is clearly demonstrated in non-pressure immersion techniques where complete impregnation can be achieved, provided sufficient time is allowed.

Diffusion treatment

Prolonged used of immersion plant is economically unrealistic but diffusion techniques can still

FIGURES 5.3 and 5.4 Diffusion treatment at a modern sawmill in Papua New Guinea. Sawn wood is immersion-treated at the end of the production line and then close stacked in sealed sheds to permit diffusion of the preservative.

be used. A concentrated water-borne preservative solution can be applied by immersion or spray to green wood possessing a high moisture content in excess of perhaps 50%. The wood is then close stacked and wrapped or placed in special sealed buildings to inhibit evaporation so that the preservative can penetrate deeply by slow diffusion (Figs 5.3 and 5.4). Timbor borate treatment is applied in this way and is perhaps the most realistic method for the treatment of spruce – complete penetration can be achieved and the relative impermeability of the spruce, following drying, means that losses through leaching are unlikely, provided that the source of moisture is not prolonged and continuous. Spruce treated in this way is not sufficiently reliable for ground contact conditions but is ideal for other situations in buildings as it provides protection against both insect attack and fungal decay caused by accidental leaks or condensation. Bifluorides are often applied similarly but much of their deep penetration can be attributed to hydrogen fluoride which is readily lost by volatilization and leaching. Potassium fluoride mixtures such as Osmose can also be applied in this way but, if fixed treatments are required, preservative fixing by ammonia or acetic acid loss are most suitable provided that

fixation is delayed by wrapping for a sufficient period to permit complete diffusion.

Spray treatment

Spray treatments are generally unsuitable for severe hazard preservation as they are only superficial and the protection can be readily damaged by the development of shakes or splits and by woodworking. However, spray treatments cannot be seriously faulted if they are applied sufficiently generously to a completed structure and they are thus perfectly adequate for remedial in-situ treatments; these are described in more detail in the book *Remedial Treatments of Buildings* by the present author. Brush treatments should never be considered for the application of preservatives as they cannot achieve loading sufficient to give even superficial protection.

Ideal preservation

The ideal situation would be to use only naturally durable wood but adequate supplies are not available and it is economically unrealistic to reject all sapwood and use only durable heartwood. An alternative ideal situation would be for all wood to be treated throughout its thickness at the mill, so that it would remain

159

reliably preserved even if subjected to extensive wood-working. This can be achieved in permeable species such as Corsican, South African and Southern pine using normal pressure impregnation techniques, even using standard copper–chromium–arsenic (CCA) preservatives. Alternatively many species, including some normally impermeable species such as spruce, can be treated by diffusion with borates to give wood that is completely reliable in relatively protected above-ground situations, including joinery (millwork) where the necessary protection is provided by a coating system.

5.2 Uses of preserved wood

Ground contact

Preservation was originally introduced as a means to avoid the deterioration that occurs when untreated wood is used in various service conditions, but the introduction of reliably preserved wood effectively introduced an entirely new structural material which can be used in new situations for which untreated wood was never previously considered – such as for durable wood house foundations. The major use of preserved wood is certainly in ground contact conditions. In many respects poles, posts and piles present similar technical problems as they all involve ground contact conditions and they vary only in their dimensions and in the fact that poles have most of their length above the ground in contrast to piles which usually have most of their length below ground.

Poles, piles and posts

Round wood transmission poles (Fig. 5.5) are used throughout the world, the principal advantages of wood being excellent strength-to-weight properties and elasticity under load. Naturally durable wood is rarely used and most poles are vacuum/pressure treated with creosote or water-borne salt preservatives. In relatively permeable woods such as Southern

FIGURE 5.5 Eucalyptus transmission poles in Australia impregnated with Tanalith C (CCA) preservative. (Hickson's Timber Products Limited)

pine complete impregnation is achieved and almost any fixed preservative is suitable; these are the Class A conditions shown in Table A.2 in Appendix A.

Where the sapwood alone is permeable but the heartwood is moderately durable, as in European redwood, resistance to movement is desirable to reduce the tendency for checks or splits to develop and expose the untreated heartwood. Water-borne preservatives are less effective than creosote, which is particularly efficient in reducing movement, although water-borne systems are efficient on species possessing low movement where there is little likelihood of the development of checks. Checking is often ignored, yet in tropical areas a wood with a large movement can fail structurally, simply due to the physical damage that results. Whilst European redwood is normally considered to possess permeable and non-durable sapwood, impermeable and moderately durable heartwood and only medium movement, these properties apply only to wood that has grown relatively slowly. With fast-grown wood the heartwood may be non-durable but it is also more permeable so that reasonable penetration can be achieved with normal vacuum/pressure processes, although it should be appreciated that where this fast-grown wood is included in a treatment charge the absorption of preservative must be increased to ensure reliable protection.

Douglas fir also possesses non-durable sapwood and durable heartwood but even the sapwood is resistant to impregnation and poles can be reliably preserved only if they are incised. Spruce possesses sapwood that is resistant to impregnation but incising has only limited advantages as the heartwood is also non-durable. However, as the wood is relatively impermeable a deposit of unfixed preservative within the heartwood possesses excellent resistance to leaching – rapidly fixed copper–chromium–arsenic (CCA) preservatives will treat the sapwood only to the depth of incising but with copper–chromium–boron (CCB) preservatives the boron component

will continue to diffuse, significantly improving the durability of the heartwood. The slow fixation preservatives which fix by ammonia or acetic acid loss are more efficient as they can combine this protracted diffusion with ultimate fixation. However, it must be appreciated that higher concentrations are required if such diffusion occurs. The wrapping of the treated wood to prevent drying and to allow for diffusion is probably best avoided for poles so that higher loadings can be achieved through relatively rapid fixation in the external sapwood zone, where there is the greatest decay risk, but slow air drying ensures that the inner moisture content changes only slowly and significant diffusion is still able to occur.

The increasing scarcity of suitable sizes and species of wood for transmission poles means that there is increasing interest in the use of more readily available species, such as relatively impermeable spruce, and in the manufacture of poles by laminating smaller sections. In laminated poles complete impregnation is essential and it is important to appreciate that a species such as European redwood with an impermeable but moderately durable heartwood is entirely unsuitable – the lamination ensures that the heartwood is exposed to ground contact which represents the greatest decay risk, whereas in round poles the heartwood is protected by reliably preserved sapwood. It may seem strange but spruce is likely to be more reliable than European redwood laminated poles because laminated spruce poles can be incised after manufacture to give a treated zone to a fixed depth and the low movement of spruce avoids the danger of the development of shakes that may penetrate through this relatively superficial treatment.

In tropical areas problems are frequently encountered in the treatment of hardwoods, particularly Eucalypts treated with water-borne preservatives. Soft rot frequently develops at the ground line, apparently because this fungus is able to invade the cell walls which have not been penetrated by the toxic components in the preservative – this is a problem of micro-distribution of the preservative within the wood and is currently the subject of extensive investigation. Class AS preservatives in Denmark are considered to be those that are particularly suitable for use in situations where there is a danger of soft rot; this generally means continuous immersion in water in the Danish context.

In the Poulain process, described in detail in Chapter 3, a relatively light overall treatment of a pole is followed by a second and more thorough treatment of the butt. This originally involved a Rüping empty-cell treatment of the entire pole with a light creosote followed by a further treatment of the butt with a heavier oil. In more recent years creosote has been used following a water-borne and sometimes unfixed treatment to give the desired additional protection at the ground line. In the Dessemond process in France, poles were first treated with copper sulphate but mercuric chloride Kyanising or zinc chloride Burnettising treatments were also used. The Card, Tetraset and other similar processes are described in detail in Chapter 4. In recent years it has often been argued that creosote butt treatment should be applied to all poles treated with water-borne preservatives but this achieves little advantage – a fixed water-borne treatment such as CCA will give reliable protection at the ground line in normal conditions and it is the exposed part of the pole that suffers from the development of checks. Butt treatments with creosote or non-toxic bitumen have an advantage only when the pole is treated with a poorly fixed preservative which is unable to withstand the ground contact conditions.

Creosote treatment (Figs 5.6 to 5.8), has the distinct disadvantage that it is dirty and likely to bleed, perhaps causing serious damage to clothing where poles are erected in areas with heavy pedestrian traffic; the problems of bleeding with empty-cell and particularly Rüping treatments has been discussed in detail in Chapter 3. Preventing bleeding only reduces the prob-

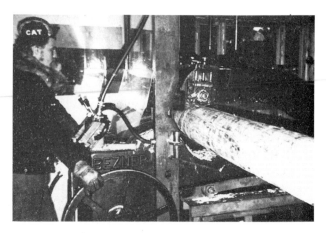

FIGURE 5.6 Peeling transmission poles prior to creosote treatment. (Industri- og Byggnadsaktiebolaget Suecia, Sweden)

FIGURE 5.7 Poles on bogies being loaded into the treatment cylinder. Creosote is heated electrically to reduce viscocity and improve penetration. (Industri- og Byggnadsaktiebolaget Suecia, Sweden)

FIGURE 5.8 The stock yard with modern handling equipment. (Industri- og Byggnadsaktiebolaget Suecia, Sweden)

lem as the poles still remain dirty. Water-borne treatments are clean and attractive but shakes or splits may develop and many of these systems contain arsenic which is a danger to livestock, even if it is proved to be reliably fixed. Fence posts can be a source of arsenic poisoning just as much as transmission poles; the easiest way to avoid arsenic toxicity is to use only arsenic-free preservatives such as the copper-chromium (ACC) types are completely reliable, except in situations where there is a serious hazard from termites or copper resistant fungi such as Poria species.

Fences receive little attention yet they represent a very large volume of treated wood. Usually, local species are used and there is thus a tendency for their value to be largely ignored as replacements can be readily obtained. However, any replacement or repair of a fence represents substantial labour costs and there is always justification for the selection of naturally durable wood or the use of a preservation process. Round posts are always best and as they have only a small section they usually consist almost entirely of sapwood. In view of the short length of the posts even relatively impermeable species can be treated by penetration through the more permeable end-grain. In many countries standard fence posts, pressure treated with creosote or water-borne preservatives, represent a normal commodity which can be readily purchased by the agricultural community, but in other countries there is a tendency to prepare posts on each individual farm. In this case preservative treatment, if it is used at all, is normally applied by the butt hot and cold method described in Chapter 3.

Although preserved construction piles are completely reliable and widely used in America, they are not popular in Europe where tubular steel and concrete piles are preferred. However, wood piles are widely used in marine situations throughout the world. Some naturally durable woods are used such as greenheart but preserved piles, particularly incised creosoted Douglas fir,

are most popular. In situations where there is a risk of marine borers it is necessary to use either very high retentions of creosote or additives such as arsenic or contact insecticides which will prevent damage by gribble. Whilst marine defence works such as groynes must be similarly protected, the superficial and decking timbers in wharfs, jetties and marinas represent only normal decay hazards and Class A preservatives are completely satisfactory for such structures.

Railway sleepers (ties)

The first use of pressure creosoted wood was for railway sleepers (ties). Wood sleepers (Fig. 5.9) are still extensively used but the declining availability of large-section wood has progressively increased their cost and metal and particularly concrete sleepers have been adopted in several countries for economic reasons, although without taking proper account of the life of the sleeper which is almost indefinite with creosote treatment to a suitable wood. In recent years the system of mounting the rails in chairs screwed to the sleeper has been abandoned in many countries in favour of the use of flat-bottomed rails secured with spikes. Unfortunately, the spikes do not hold so well in creosoted wood and there are several disadvantages with water-borne treated wood such as movement splits, exposing only moderately durable heartwood in European redwood sleepers, and electrical insulation problems with salt systems, where signalling systems operate through the rails. With new high-speed tracks, where spiked flat-bed rails are unsuitable, and where there are doubts about the life of concrete sleepers, there is now a tendency to return to creosoted wood sleepers.

With European redwood transmission poles and sleepers there is a danger, particularly with water-borne treatments, that untreated moderately durable heartwood will be exposed by cracking and will slowly decay. When such cracking is observed in adequate time it is possible to carry out remedial treatments using

FIGURE 5.9 Using preservative cartridges of Wolmanit TSK to protect heartwood at risk in sleepers (ties) through checks extending through the outer preserved zone. (Dr Wolman GmbH)

the Cobra and similar injection processes described in Chapters 3 and 4. Diufix, a spreadable mixture of creosote, tar, pitch and filler, was developed for coating railway sleepers to fill existing cracks and reduce the tendency for further shakes to develop.

With some preservatives a further problem is slow and progressive Soft rot attack at the ground line. With softwood poles this damage generally occurs only with the old fluorine–chromium–arsenic–dinitrophenol (FCAP) treatments; with copper–chromium (ACC) and copper–chromium–arsenic (CCA) preservatives it occurs only on hardwoods. In 1928 Allgemeine Holzimprägnierung GmbH introduced AHIG, the first pole bandage for wrapping round the exposed ground line of a pole to control Soft rot damage (see Fig. 3.5). AHIG, which consists of a water-proof bandage lined with Wolman salts, has since been followed by many similar products such as the Osmose

bandage and Pile Gard. In the Mayerl process a trough is fixed round the pole at the ground line and filled with creosote which is then slowly absorbed into the pole. These processes are specific remedies for progressive surface Soft rot and injection is essential if an attempt is to be made to control heart rot developing in untreated heartwood.

Road works

Wood blocks treated with creosote and tar were extensively used in the past as road paving blocks. They were laid with the end-grain upwards and gave an exceptionally durable and resilient road. Whilst they are no longer generally used for public roads they are still used in parts of Europe as flooring in heavy industrial works (Fig. 5.10). In modern road building, preserved wood is most extensively used for fencing and for crash-barrier posts – large-section wood posts can be installed directly into

FIGURE 5.10 End-grain wood blocks impregnated with creosote and used as a heavy industrial floor. (Orben Bois SA)

the soil whereas steel posts must be mounted in concrete if they are to provide adequate resistance to impact damage. Both fencing and crash-barrier posts are usually treated with water-borne preservatives, particularly copper–chromium–arsenic (CCA) types at the Class A retentions shown in Table A.2 in Appendix A.

Bridges

Bridges are also sometimes constructed from wood but it is important to be aware of the dangers of exposing heartwood which is only moderately durable when, for example, sawn European redwood is employed. Incised European whitewood or spruce is more reliable, even though it is classified as non-durable and impermeable, than European redwood or Scots pine with easily penetrated sapwood and moderately durable heartwood; the low movement of spruce means that incising results in treatment to a controlled depth which is unlikely to be penetrated by the development of shakes.

Buildings

Many exposed structures represent problems similar to those for bridges but in buildings (Figs 5.11 to 5.14) there is usually less risk as they should be designed to ensure that wood remains dry. For this reason a special Class B is shown in Table A.1 in Appendix A for buildings, including cladding and structural elements exposed to the weather, as above-ground conditions are generally less severe than ground contact. Wood preservation is required in flat roofs, swimming pool roof linings and industrial buildings where a decay risk may arise through condensation, but in other circumstances the main reason for a preservative treatment may simply be the desire to guard against future damage by accidental leaks or by House Longhorn beetle in temperate areas and Dry Wood termites in the tropics. Where insecticidal protection is required, preservatives meeting the Class I requirements in Table A.1 are required. Generally, water-borne preservatives containing

165

FIGURE 5.11 Pine, pressure impregnated with an organic solvent preservative BP Hylosan, used as cladding for a prize-winning housing project in Sweden. (Svenska BP Aktiebolag)

FIGURE 5.13 Pine, impregnated with BP Hylosan, used as ceiling and wall lining to a swimming pool in Sweden. (Svenska BP Aktiebolag)

FIGURE 5.12 Preservative-treated framing and Douglas fir plywood used in Canada for construction of durable house foundations, a system widely used in North America. (Council for Forest Industries of British Columbia)

FIGURE 5.14 Pine, impregnated with Boliden K33 preservative used as a durable cladding for a factory building in Sweden. (Anticimexbolagen)

arsenic or boron are most economic but these are stomach poisons and tasting damage may be significant, particularly in the case of exposure to termites. The use of contact insecticides such as Lindane, Dieldrin or Permethrin is desirable to avoid the damage but their effective life is considerably less. With non-flying termites, damage is normally avoided by poisoning the soil around the building or by using shields which will prevent the termites from gaining access; these are illustrated in Fig. 5.15 and are rather reminiscent of the staddle stones used in old English barns to prevent rats from gaining access to the stored grain.

Fencing

Whilst wood used above ground level in buildings generally represents a considerably reduced risk compared with that in ground contact conditions, there are situations where wood is exposed to the weather and may decay where rainwater can penetrate into joints or cracks. It is perhaps worth mentioning that one example of this risk occurs in fencing where the mortices used in the construction of gates and for fixing rails to posts represent water traps and are therefore invariably the areas where decay progresses most rapidly, even if naturally durable wood such as oak is employed.

Joinery (millwork)

In buildings, the external cladding does not usually present any serious problems and the main danger is associated with the joinery (millwork) such as the window and door frames. Decay develops under the paint or varnish coating when water is trapped following absorption through splits at the joints. The natural reaction is to introduce a preservative to prevent decay but splits then continue to occur at the joints and water is absorbed which, even if it is unable to cause decay, results in preferential wetting failure of the paint coating. The use of water-repellent organic-solvent preservatives tends to prevent the water from being absorbed, although cracks at the joints still occur so that water repellents tend to delay rather than completely prevent failure. The best way to reduce cracks is to use only wood with low movement and natural durability; suitable

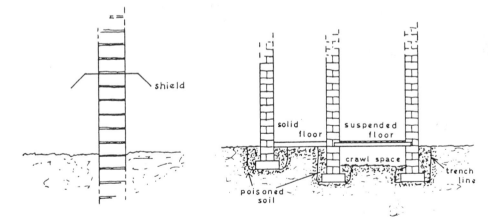

FIGURE 5.15 Protecting buildings from non-flying termites. Shields (or caps) on supporting walls and all pipes and cables passing from the oversite to the floor will give some protection but regular inspection of the crawl space is necessary and any mounds or tunnels found, which tend to by-pass the shields, must be destroyed. Soil poisoning is more reliable and now more commonly used. Trenches are dug, then the oversite is levelled and the site sprayed. Treatment is also applied to all fill returned to the trenches. In this way the building is isolated by a poisonous zone.

tropical hardwoods are available. Alternatively, preservative treatment can be used but it is still essential to use wood with low movement to avoid the splitting of the coating system at the joints so that a longer decorative life can be achieved.

Remedial treatments

Remedial treatments in buildings (Fig. 5.16) are described briefly in Chapters 3 and 4, but these are extremely complex processes which are also related to dampness problems and masonry deterioration and they are best considered entirely separately, as in the book *Remedial Treatments of Buildings* by the present author. The economics of preservation versus remedial treatments in buildings must also be considered – is it more sensible to take precautions during construction or simply to remedy defects if they develop? In ground contact conditions it is clear

that the use of naturally durable or adequately preserved wood is essential. European redwood or Scots pine joinery (millwork) is almost certain to decay, despite regular painting, and preservation is justified. Where there is a risk of House Longhorn beetle preservation is again justified for the carcassing or framing components of a building but it cannot be justified where there is a risk of attack only by Common Furniture beetle which can do less damage to the structure. Sometimes preservation is applied as there may be a slight danger of an accidental leak but it is unlikely that it can be justified, although if there is known to be a condesation problem, as in flat roof construction, precautions preservation is essential.

The conclusion that must be reached is that there are many situations in buildings where preservation is essential or clearly desirable, but it is equally apparent that in many countries

FIGURE 5.16 A water-in-oil emulsion Wykemulsion being used for remedial treatment of a roof. Organic solvent preservatives are often preferred but this emulsion reduces the fire risk. (Cementone-Beaver Limited)

the need for preservation is largely ignored. The proposal that treatments should be introduced in under-developed countries is particularly interesting; at present a large proportion of the national effort is devoted to replacing decayed buildings and even a preservation treatment of limited efficacy may result in a substantial improvement in national prosperity.

Boats

In boats (Fig. 5.17) it is usual to use only naturally durable wood to avoid fungal decay. The main danger arises through penetration of rain through decking and upper works together with poor ventilation within the hull. There is no reason why properly preserved wood should not be used as an alternative to naturally durable wood, a fact which is generally accepted in the case of plywood, but the use of wood in boat building has progressively declined in recent years, largely as a result of the unreliable performance of wooden boats. This can be attributed partly to the serious decay that occurred regularly some years ago in non-durable core veneers in plywood. Whilst non-durable veneers are no longer permitted, plywood still possesses a poor reputation. The other problem with wooden boats is the need to paint them regularly; this could be avoided by the use of treatments which would prevent preferential wetting failure. These problems are described in more detail in Chapters 3 and 4. There are opportunities for preservation treatments to be used more extensively but it is first necessary to re-establish the good reputation of wood in boat building. One serious problem concerns the attitude of paint manufacturers; they are reluctant to adopt systems that will reduce paint failure as they are largely dependent on maintenance painting for their turnover, although this short-sighted attitude has resulted simply in the abandonment of painted boats and is the main reason for the substantial reduction in the use of wood in boat building.

FIGURE 5.17 Pine, pressure treated with Boliden K33 salt preservative, used for piles, beams and decking of a yacht jetty. (Anticimexbolagen)

FIGURE 5.18 A simple pole barn constructed from wood impregnated with Tanalith C (CCA) preservative. Hickson's Timber Products Ltd)

FIGURE 5.19 Silos in Sweden contructed from pine impregnated with Boliden K33. (Anticimexbolagen)

Agriculture

Agricultural buildings (Figs 5.18 and 5.19) are often constructed from wood in all parts of the world. In recent years there has been a tendency to abandon the use of frame buildings and to adopt pole construction instead, an example being the pole barn where pressure-treated roundwood is used for the basic structural members. The use of preserved clapboard

cladding enables this form of construction to be adopted for other purposes such as cow-houses and implement sheds. There are a multitude of small uses for preserved wood in agriculture and horticulture, a few being the construction of greenhouses, mushroom boxes, stakes for fruit trees and vines, and fences and gates.

Mines

In mines the dampness and constant temperature conditions favour decay. Two basic types of support are required, the working-face supports or props and the linings to the main roadways. Wood is still extensively used for props but it is usually considered that there is little justification for preservation as only a relatively short life is required. In the case of the more permanent roadways there is a tendency for steel to be adopted but wooden boards are often placed between steel arches and these must be treated. Usually, water-borne preservatives are used, particularly the copper–chromium–arsenic (CCA) types, and there is also a demand for fireproofing treatments for wood used in principal roadways, although it is not clear that wood necessarily contributes to fires in mines as it is normally too damp.

Australian quarantine requirements

Two special preservation requirements must be mentioned. The first concerns the Australian Quarantine Regulations which require that all imported wood forming part of disposable or re-usable packaging (cargo containers) must be treated to ensure that it does not introduce a wood-borer risk. The reasons for these regulations have been discussed in detail in Chapter 2 and suitable Class 1 type preservative retentions are shown in Table A.1 in Appendix A.

Cooling towers

Another special use for preserved wood is in cooling towers where the fill slats are exposed to hot water and are thus particularly susceptible to Soft rot degradation. Whilst there have been experiments with a number of alternative materials such as glass, asbestos and concrete, the low density of wood makes it particularly attractive for this purpose and it is entirely reliable if treated with a suitable preservative such as a copper–chromium–arsenic (CCA) system.

5.3 Health and the environment

In recent years the health and environmental dangers associated with wood preservation have attracted particular attention. Restrictions on the use of existing preservatives and the requirements for approval of new preservatives have become increasingly stringent and are now causing serious difficulty to the industry. These changes have not necessarily resulted in reduced risks to health and the environment, as the development of safer preservative systems is now discouraged by the costs involved in submitting new preservatives for approval and it has been necessary, for economic necessity, to extend the life of established preservative systems which would not be acceptable if they were submitted for safety approval today.

All wood preservatives contain toxic components but there is no justification for their prohibition. Regulations should specify the precautions that are necessary to ensure their safe use in terms of the hazards to operatives during formulation and use, to the users of treated wood, and to the environment. In some cases these precautions may mean that it is uneconomic to use a particular product and realistic control is therefore achieved. For example, arsenic compounds are very toxic but they can be safely used in wood preservation with appropriate strict controls on the handling of treating solutions, but modern preservative formulations ensure that the arsenic is ultimately fixed in the wood so that it cannot easily affect users or the environment by leaching or volatilization. Control is much easier at treatment plants than during the handling and working of treated wood or service in a structure, remedial wood treatment in buildings presenting the greatest risks so that it is discussed later in detail, although solution spills and wind-blown dust from treatement plants can cause problems; dust dangers are not significant with most preservatives as powder formulations have been largely replaced by pastes and solutions.

Arsenic

The arsenic content in CCA, FCAP, CAA, ZAA, ACA, ACZA and similar formulations is a serious problem. it is unfortunately true that cattle have been poisoned as a result of licking treated transmission poles and fence posts but this normally occurs only with preservatives in which the arsenic is not fully fixed, such as K-33 which has now been largely withdrawn, and only in areas where there is a natural salt deficiency – the danger can be completely avoided by providing proper salt licks and using only highly fixed preservatives. Arsenic preservatives are banned in buildings in some countries such as Finland, yet in others they are used even for the treatment of playground equipment. In Switzerland, arsenic preservatives are banned for the treatment of transmission poles as they may introduce environmental pollution. In most countries these hazards are considered to be insignificant with CCA preservatives in view of their excellent fixation, although fixation is only reliable with CCA-C and BS 4072 formulations, and the main fears are related to the possible volatilization of arsenic when treated wood is destroyed by burning.

Yet another fear is the danger of arsine poisoning. In about 1890 many fatalities occurred in homes and these were eventually attributed to arsine poisoning. Wallpaper decorated with arsenical dyes had been attacked by the fungus *Scopulariopsis brevicaulis*. at that time known as *Penicillium brevicaule*; it is now known that other fungi can generate arsine in this way. There is no danger of arsine poisoning from preserved wood provided that the fungicidal components are able to control the fungi that generate arsine. If preservation is required against insect attack alone, as for the treatment of wood in accordance with the Australian Quarantine Regulations or for Furniture Beetle control in New Zealand, there is clearly a temptation to use just a simple arsenical preservative and there is then a danger of arsine poisoning if fungi are able to develop. Many insects are dependent on or encouraged by the presence of fungal attack and fungicidal components in multi-component formulations therefore assist in their control. This factor is completely ignored in Australia and New Zealand where the minimum retentions of approved preservatives are based solely on an arsenic retention of about $0.97\,kg/m^3$ As_2O_5.

There have been fears, particularly in the United States, concerning the carcinogenic dangers associated with the arsenic contents in wood preservatives. There have been extensive enquiries and the current evidence suggests that the carcinogenic properties are largely associated with arsenic trioxide, As_2O_3, and arsenites rather than with the arsenic pentoxide, As_2O_5, and the arsenates that are used in modern preservative formulations. The carcinogenic dangers associated with arsenical wood preservatives are slight but arsenic pentoxide is prepared from the trioxide and increasing controls on the latter compound are likely to introduce manufacturing difficulties, perhaps leading to scarcity and increased cost. The dangers associated with fixed arsenic wood preservatives are often exaggerated, sometimes by manufacturers of competitive products; the enquiries in the United States were prompted by the cement and concrete industry which feared the competition of pressure-treated wood foundations in domestic construction!

Chromium

Arsenic dangers always attract most attention but chromium may represent the greatest hazard in wood preservation. If properly formulated preservatives are used in a competent and responsible manner the dangers are very slight and this is confirmed by the very low level of illness or injury in this long-established industry, but it must be recognized that dangers exist. Tropical conditions discourage the use of protective clothing and operatives then frequently suffer from chrome ulcers which are painful and difficult to heal but it is interesting to note that there are no arsenic problems, and the injuries sustained are an indication of poor plant hygiene and control rather than serious criticism of the preservative formulations involved.

Creosote

In recent years health risks associated with the use of creosote have been more closely scrutinized, the main risks being identified as the reported carcinogenic properties of polycyclic aromatic hydrocarbons, including the benzopyrenes in creosote. The benzopyrene content can be limited by using only creosote distilling at higher temperatures, a restriction that does not affect impregnation grades but which is now restricting the availability of lighter creosote for surface application, for maintenance of fencing. In other respects the use of creosote does not present any unusual health hazards, provided that plants are operated with proper care and persons handling creosote and treated wood observe normal personal hygiene precautions.

Chlorophenols

In the United States the wood preservation industry still uses about 20 000 tonnes of pentachlorophenol annually despite concern regarding possible health hazards. Whilst it is certainly true that chlorophenols are toxic, this is a problem that applies to all wood preservatives and they should therefore be handled with care. Some fatalities have occurred but these have all been associated with abnormal absorptions of pentachlorophenol through a failure to take sensible precautions. There were several fatalities in France when operatives who were stripped to the waist in very hot weather became heavily contaminated with sapstain treatment, and several fatalities occurred in Germany in a similar way through people bathing in water contaminated with pentachlorophenol remedial treatment. There are also fears that the dioxin impurities in chlorophenols may be particularly hazardous but, whilst there may be distinct dangers associated with the herbicide 2,4,5-T, which is a trichlorophenyl acetate, there is no evidence of similar dangers associated with pentachlorophenol or even the trichlorophenol with which it is sometimes contaminated. Apparently, this is because the latter is 2,4,6-trichlorophenol and generates a different range of dioxin impurities which are actually less toxic than the chlorophenols from which they are derived. Environmental theoreticians have suggested that pentachlorophenol should be replaced in stain control by less persistent trichlorophenol but this compound is less effective and must be used at higher concentrations, and the dioxin impurities tend to be more toxic than those associated with pentachlorophenol, so that the continued use of pentachlorophenol would appear to be best in terms of both handling safety and environmental protection. Tetrachlorophenol is also sometimes used and represents an intermediate risk.

Chlorinated hydrocarbon insecticides

The chlorinated hydrocarbon insecticides DDT, Dieldrin and Lindane have attracted particular attention at times because of their interference in environmental food chains when used in agriculture or horticulture. Wood preservation treatments do not normally interfere in the environment in this way because wood preservation insecticides are designed to achieve complete control of borers within the treated wood. In contrast, in agriculture and horticulture, killed insects and treated survivors may be eaten and enter food chains. Lindane is still permitted for many wood preservation purposes but much safer pyrethroid insecticides have been introduced and are now preferred, particularly Permethrin; various natural and synthetic pyrethroids have been approved for many years for use in the food industries for insect control.

Organotins

The organotin compound tri-n-butyltin oxide has been in use in wood preservation since about 1960 without any serious health problems being reported, other than dermal and respiratory irritation which can be attributed to poor operative technique and inadequate personal hygiene, although other organotin compounds are exceedingly toxic. This is a further example of the situation that has been encountered in relation to chlorophenol dioxins – although there may be some exceedingly toxic compounds within a group, that does not mean that all compounds in that group are similarly toxic and some may be virtually non-toxic.

Copper and zinc soaps

Copper and zinc naphthenates have been used for many years in the wood preservation industry without reports of unusual health problems, although the naphthenic acid liberated from these treatments has a distinct musty pungent odour which is unpleasant and irritating. A leading manufacturer of preservatives of this type has replaced the napthenates in recent years with acypetacs compounds which avoid this musty odour, although they produce instead a slight sickly odour to which some individuals seem to be particularly sensitive, and which has caused difficulty in houses where excessive preservative has been applied.

Remedial treatments

Clearly, remedial treatment contractors have a duty to ensure the good health and safety of their own operatives as well as the occupiers of treated buildings. Proper protective clothing must be provided, although it is difficult to

173

ensure that it is worn at all times. For example, during the spring roof spaces may become rather warm in contrast to the winter period and operatives may be tempted to remove clothing, perhaps during spraying. The operative then leaves the roof and perhaps sits in the sun without a shirt or vest and there is then a danger of mild sunburn, accompanied by considerable irritation if the skin is affected by organic solvents. Nose bleeding can also occur when some preservative vapours are encountered in high temperature conditions, but it must be appreciated that some operatives are more sensitive than others, and particularly sensitive persons should never be employed for this type of work. Masks and barrier creams are often recommended but neither is really advisable. Simple gauze masks tend to absorb treatment fluids, perhaps exposing the user to abnormal concentrations of toxicant vapours; without a mask, the operative would probably take more care to avoid unnecessary breathing of preservative spray or vapour. Barrier creams can also give operatives unjustified confidence and it is far better to train operatives to take the necessary care. Obviously all operative gangs must be aware of the health and fire dangers, and must be aware of the action that should be taken in an emergency; a first aid kit and fire extinguishers should always be available to remedial treatment operatives.

There are several important points that should be borne in mind when carrying out remedial treatment in buildings using organic solvent preservatives. Low-pressure sprays should be used with coarse jets to ensure that the maximum volume can be applied to the timber surface without the excessive volatilization of solvents that occurs if high pressures and fine jets or air-entrained paint sprays are used. Whilst it is essential to achieve the maximum loading of preservative on the wood to ensure maximum penetration, dripping of excess fluid must be avoided and care must be taken to ensure that electrical cables are not treated unneces-

sarily and preservative does not enter junction boxes or other electrical fittings. Treated areas must be freely ventilated to disperse solvent vapour which is a fire hazard and which may affect electrical cables and cause staining around ceiling roses and wall switches. Electrical installations in the treated areas should be disconnected during treatment and even for several days afterwards as there is a danger that sparks may ignite solvent vapour. Smoking, naked lights and plumbing activities must be prohibited in the area for 7 to 14 days depending on the nature of the solvents involved, and notices to this effect should be posted at the entrances of the property and at the entrances to roof spaces and other treated areas. Insulation materials must always be lifted before treatment and replaced later after the solvent has completely dispersed, certainly not less than seven days after treatment in any circumstances, and insulation must never be sprayed wih preservative as there is then a real danger of spontaneous combustion.

Some of the phenolic preservatives, particularly the chlorophenols, can cause treatment operatives severe respiratory and dermal irritation, particularly if excessive spray pressures are used which result in preservative atomization and spray drift. Dermal irritation problems are often due to the solvents alone and enquiries usually disclose that the individuals are sensitive to similar solvents such as gasoline, kerosene, white spirit and turpentine; such problems are usually associated with fair skin and are aggravated by exposure to sunlight. This dermal irritation is also aggravated by some preservative biocides, particularly chlorophenols such as pentachlorophenol (PCP) and organotin compounds such as tri-n-butyltin oxide (TBTO), although sensitivity to these biocides varies enormously and, if normal precautions are observed in use, problems are only encountered by particularly sensitive individuals. In extreme cases, respiratory irritation can occur and cause coughing and bleeding from the nose but such reactions are usually related to extreme exposure such as spraying preservatives in roof spaces during very hot weather; these problems can be reduced by using coarse low pressure spray application to flood the surface of the wood with preservative which can then be absorbed by capillarity, avoiding high pressure sprays which cause atomization and rapid volatilization, but improved ventilation may also be necessary.

If sufficient ventilation is provided following treatment the solvents will rapidly disperse, leaving only preservative deposits of low volatility which do not normally cause persistant odours. There have been various suggestions in recent years that these treatments are dangerous to health but treatments must be approved by the health and safety authorities in most countries. The health risks associated with current preservatives have therefore been carefully assessed and there is no reason to suppose that they present significant risks, either to treatment operatives or to persons resident in treated buildings. Obviously, treatment operatives are severely exposed and would be expected to suffer most seriously from any health hazards but, although there are perhaps 5000 to 10 000 operatives employed in the remedial timber treatment in the British Isles, reports of health problems are very few indeed, despite the fact that most operatives work within the industry for many years; on the contrary it seems that operatives suffer less from some common illnesses such as colds and influenza!

There were several proprietary remedial treatment preservatives some years ago which were based on o-dichlorobenzene or on mono- or dichloronaphthalene. These biocides are oils which can be readily absorbed through the skin and they certainly presented a danger of liver damage to treatment operatives. The dangers were much less with the solid polychloronaphthalene waxes which could not be absorbed in this way and no illnesses were reported despite their extensive use at very high concentrations over many years. Pentachlorophenol attracted attention in the past because of its pungent and irritating odour when applied, but in recent years attention has concentrated on the dioxin impurities that may be present in chlorophenols, as described earlier in this section.

Health problems are not confined, of course, to remedial treatment wood preservation but obviously spraying in confined spaces represents the most intensive exposure that is likely to be encountered. The other common problems that arise are dermal and respiratory problems due to handling timber treated with organic solvent preservatives, particularly in hot weather, and careless operation of treatment plants involving preservative spillage, vacuum pumps discharging into working spaces, and timber dried after treatment in working spaces, all problems that are associated with careless handling rather than any defect in the preservative system. However, there are periods when a series of complaints arise, apparently associated with a particular preservative biocide. Obviously, reports of problems can prompt further unjustifiable complaints, but a series of incidents usually have some common cause which is often very difficult to identify. Some of the complaints in recent years seem to be associated with Lindane treatments and others with TBTO treatments. In both cases unusual volatility seems to be involved which sometimes affects treatment operatives but which is also readily apparent to the occupiers of treated buildings. In such cases it must be suspected that the Lindane and TBTO were poor quality products containing impurities which have caused these problems as such problems are not associated with the pure compounds. Lindane is defined as 99% pure γ-isomer of hexachlorocyclohexane, previously known as gamma-benzenehexachloride, and it seems that some material contains much higher concentrations of other isomers. TBTO often contains a stabilizer, and some stabilizer compounds interfere with the fixation of the TBTO to the treated wood so that the TBTO remains volatile. Such problems may be indications of inadequate quality control but they are very rare; when complaints are investigated it is generally discovered that normal precautions have not been observed, particularly in relation to excessive treatment levels and ventilation following treatment.

Further reading

Wood preservation is continuously developing in response to new research findings and economic and safety pressures. Any list for suggested further reading would be largely obsolete and it is therefore more appropriate to give guidance on sources of further information.

Guidance on current approval requirements can be obtained from national organizations responsible for standards and regulations.

Any reader of this book can obtain current information on wood preservation from papers presented during the proceedings of the following organisations:

The International Research Group on Wood
  Preservation,
Box 5607,
S-114 86 Stockholm,
Sweden.

American Wood Preservers' Association,
P.O. Box 849,
Stevensville,
MD 21666,
U.S.A.

British Wood Preserving and Damp-proofing
  Association,
Building 6,
The Office Village,
4 Romford Road,
Stratford,
London,
E15 4EA,
England.

The reference lists at the ends of the papers will suggest further reading.

TABLE A.1 Typical preservative retentions for Baltic redwood*

Treatment	Sapwood retentions (kg/m^3)*			
	Class A (Ground)	Class B (Buildings)	Class M (Marine)	Class I (Insects)
Tar oil				
Creosote[1]	135[1]	–	270[1]	–[1]
Inorganic				
Copper–chromium (ACC)				
$CuO + CrO_3$	9	9	–	–
Celcure (Celcure N, Tancas CC)	18	18	–	–
Copper–chromium–arsenic (CCA)				
$CuO + CrO_3 + As_2O_5$	9	9	18	–
As_2O_5	–	–	–	0.97
BS 4072, type 1 (Celcure A)	15	15	30	4.8
type 2 (Tanalith C. Tancas C)	15	15	30	5.6
K33 paste (Boliden K33, Lahontuho K33)	12	12	24	2.9
Tanalith CCA	12	12	24	(4.5)
Copper–chromium–boron (CCB)[2]				
$CuO + CrO_3$ (as for ACC)	9	9	–	–
H_3BO_3	–	–	–	5.3
Wolmanit CB, Tanalith CBC, Celcure M	21	16	–	16
Copper–chromium–fluorine (CCF)				
$CuO + CrO_3$ (as for ACC)	9	9	–	–
Basilit CFK	15	15	–	–
Copper–chromium–phosphorus (CCP)[3]				
$CuO + CrO_3 + P_2O_5$	9	9	–	–
Boliden P50	18	18	–	–
Fluorine–chromium–arsenic–phenol (FCAP)				
$F + CrO_3 + As_2O_5 + DNP$	–	12	–	–
As_2O_5	–	–	–	0.97
Tanalith U, Wolmanit FCAP, Wolmanit UA, Basilit UA	–	15	–	6.2
Copper–chlorophenol[4]				
KP-Cuprinol; K-salt + P-salt	21	21	–	(21)[4]
Copper–caprylic acid[4]				
Cuprinol Tryck (pressure)	40	40	–	(40)[4]
Organotin				
Permapruf T, BP Hylosan PT	(10)	(10)	–	(10)
Borate				
Timbor (diffusion or pressure); H_3BO_3	–	5.3	–	(5.3)
Organic				
Pentachlorophenol[5]	–	6.4	–	(5)
Metal naphthenate; Cu	–	3	–	–
Zn	–	4	–	–
Tri-*n*-butyltin oxide	–	1.3	–	(3)
Lindane, γ-HCH	–	–	–	1.6
Dieldrin, HEOD	–	–	–	0.8
Permethrin, NRDC 143	–	–	–	0.3

Further reading

Wood preservation is continuously developing in response to new research findings and economic and safety pressures. Any list for suggested further reading would be largely obsolete and it is therefore more appropriate to give guidance on sources of further information.

Guidance on current approval requirements can be obtained from national organizations responsible for standards and regulations.

Any reader of this book can obtain current information on wood preservation from papers presented during the proceedings of the following organisations:

The International Research Group on Wood
    Preservation,
Box 5607,
S-114 86 Stockholm,
Sweden.

American Wood Preservers' Association,
P.O. Box 849,
Stevensville,
MD 21666,
U.S.A.

British Wood Preserving and Damp-proofing
    Association,
Building 6,
The Office Village,
4 Romford Road,
Stratford,
London,
E15 4EA,
England.

The reference lists at the ends of the papers will suggest further reading.

Selection of a preservation system

Decay hazard

Throughout the world the most important risk situation involves the use of wood in ground contact, where there is a severe danger of fungal decay, particularly by Basidiomycetes. These fungi are generally unable to develop when wood is immersed in water, although there is then a danger of Soft rot and borer attack in marine situations. Above ground level there is still a danger of decay where rainwater may remain trapped in joints or splits in wood exposed to the weather, and this danger can be considerably heightened where a split occurs through a relatively impermeable coating such as paint or varnish, which will tend to restrict the dispersion of the water by evaporation. In structures, the framing or carcassing components are generally protected from the weather and there is no risk of decay provided care is taken to avoid leaks and the structure is designed to avoid condensation.

Borer hazard

In most temperate areas the sapwood of many hardwoods and softwoods is susceptible to Common Furniture beetle infestation but a much greater structural risk occurs in temperate and tropical areas where there is a danger of infestation by the House Longhorn beetle and Dry Wood termites. Many termites are, however, unable to fly and they are more readily controlled by soil poisoning around a structure or the installation of termite shields rather than by treatment of the susceptible wood.

Permanence – penetration

In all preservation processes there is a need for permanence. Thus all preservatives must have adequate resistance to volatilization and oxidation, depending on their particular conditions of use, and also resistance to leaching if they are likely to be exposed to a high moisture content in ground contact conditions or in marine environments. Whilst it may appear adequate to enclose wood in a superficial envelope of protective treatment it must be appreciated that deeper penetration significantly improves the life of treatments which are slightly volatile or water soluble, and it must also be appreciated that natural splits or accidental damage may penetrate through a superficial treatment, perhaps permitting internal decay to occur. All the treatments considered in this appendix must be applied by a technique that will ensure deep impregnation, usually one which involves the use of vacuum and pressure cycles in closed cylinders.

Wood properties

Table A.1 gives a schedule of some of the most widely used preservative systems and typical

TABLE A.1 Typical preservative retentions for Baltic redwood*

Treatment	Sapwood retentions (kg/m³)*			
	Class A (Ground)	Class B (Buildings)	Class M (Marine)	Class I (Insects)
Tar oil				
Creosote[1]	135[1]	–	270[1]	–[1]
Inorganic				
Copper–chromium (ACC)				
CuO + CrO₃	9	9	–	–
Celcure (Celcure N, Tancas CC)	18	18	–	–
Copper–chromium–arsenic (CCA)				
CuO + CrO₃ + As₂O₅	9	9	18	–
As₂O₅	–	–	–	0.97
BS 4072, type 1 (Celcure A)	15	15	30	4.8
type 2 (Tanalith C. Tancas C)	15	15	30	5.6
K33 paste (Boliden K33, Lahontuho K33)	12	12	24	2.9
Tanalith CCA	12	12	24	(4.5)
Copper–chromium–boron (CCB)[2]				
CuO + CrO₃ (as for ACC)	9	9	–	–
H₃BO₃	–	–	–	5.3
Wolmanit CB, Tanalith CBC, Celcure M	21	16	–	16
Copper–chromium–fluorine (CCF)				
CuO + CrO₃ (as for ACC)	9	9	–	–
Basilit CFK	15	15	–	–
Copper–chromium–phosphorus (CCP)[3]				
CuO + CrO₃ + P₂O₅	9	9	–	–
Boliden P50	18	18	–	–
Fluorine–chromium–arsenic–phenol (FCAP)				
F + CrO₃ + As₂O₅ + DNP	–	12	–	–
As₂O₅	–	–	–	0.97
Tanalith U, Wolmanit FCAP, Wolmanit UA, Basilit UA	–	15	–	6.2
Copper–chlorophenol[4]				
KP-Cuprinol; K-salt + P-salt	21	21	–	(21)[4]
Copper–caprylic acid[4]				
Cuprinol Tryck (pressure)	40	40	–	(40)[4]
Organotin				
Permapruf T, BP Hylosan PT	(10)	(10)	–	(10)
Borate				
Timbor (diffusion or pressure); H₃BO₃	–	5.3	–	(5.3)
Organic				
Pentachlorophenol[5]	–	6.4	–	(5)
Metal naphthenate; Cu	–	3	–	–
Zn	–	4	–	–
Tri-*n*-butyltin oxide	–	1.3	–	(3)
Lindane, γ-HCH	–	–	–	1.6
Dieldrin, HEOD	–	–	–	0.8
Permethrin, NRDC 143	–	–	–	0.3

TABLE A.1 (continued)

Notes

* The retentions for Baltic redwood or Scots pine also apply to other softwoods. Hardwoods often possess naturally durable heartwood which is used without treatment but, where treatment is required, special retentions may be necessary as mentioned in the text. Retentions in kg/m^3 divided by 16 give retentions in lb/ft^3. The heartwood in Baltic redwood is relatively impermeable and moderately durable so that only the sapwood is normally treated. Sawn Baltic redwood normally contains 40–50% sapwood so that sapwood retentions are a little more than double overall retentions. In species with non-durable heartwood these quoted retentions are desirable throughout. The classification of service conditions is not standard but is similar to the Nordic system. Class A represents wood in normal ground contact conditions, such as transmission poles, fence posts, railway sleepers (ties), piles and structural foundations. Class B represents building and construction wood not in ground contact but subject to a moderate risk of decay, such as carcassing (framing), joinery (millwork) and cladding. Class M applies only when there is a risk of marine borer attack, particularly by gribble, *Limnoria* spp. Class I applies only when there is a risk of insect attack, particularly by House Longhorn beetle, *Hylotrupes bajulus*, and Dry Wood termites; preservatives meeting this class conform with the Australian quarantine requirements, but those in brackets are not formally approved.

[1] Higher creosote retentions are specified in some countries. The higher retention for Class M is unnecessary if the creosote is fortified with arsenic, Lindane, Dieldrin or Permethrin to give improved resistance to gribble. This fortified creosote should also be used where protection against termites is required.

[2] The boron content is not considered to contribute to the preservative action, but its ability to penetrate deeply is valuable on species possessing non-durable heartwood and low permeability.

[3] The phosphorus content is considered to contribute to the preservative action, although its value is doubtful in some circumstances.

[4] The insecticidal action must be considered doubtful.

[5] Retentions for classes A and B are based on the American Wood Preservers' Association standards; only $5 kg/m^3$ are required by the Nordic Wood Preservation Council.

recommended retentions on European redwood or Scots pine. These retentions also apply on most other softwoods, although actual preservation processes may vary depending upon the properties of an individual wood species, as shown in Table A.2. For example, Corsican pine, *Pinus nigra*, South African pine, *Pinus patula*, and Shortleaf of Southern pine, *Pinus echinata*, all possess non-durable heartwood and sapwood. However, both heartwood and sapwood are also permeable and these species can therefore be readily preserved, but it will be appreciated that the sapwood retentions shown in Table A.1 must be achieved throughout in such woods. European redwood or Scots pine, *Pinus sylvestris*, possesses rather similar properties when fast grown, but in slow-grown wood the heartwood is much less permeable and almost completely resistant to penetration. This slow-grown heartwood also possesses moderate natural durability and it is therefore generally considered that adequate protection can be achieved by treating the permeable sapwood alone. In Douglas fir, *Pseudotsuga menziesii*, the heartwood again has good natural durability but the non-durable sapwood is also resistant

to impregnation. However, adequate penetration can be achieved if the sapwood is incised to give access to routes for tangential penetration. Unfortunately, the same principle is less effective on European spruce, *Picea abies* – although incising enables the sapwood to be treated the heartwood is non-durable, and there is thus a danger that the outer sapwood could split through shrinkage following periodic wetting and drying, exposing unprotected and non-durable heartwood.

Spruce

In fact, however, the development of splits is not very common with spruce as it possesses low movement, as shown in Table A.2, and it is therefore more suitable for superficial treatment than woods with medium or large movement. If deep penetration into spruce can be achieved with a non-fixed preservative, such as the boron component in a copper–chromium–boron (CCB) preservative or a simple borate preservative applied by diffusion, leaching is unlikely once initial drying has occurred, in view of the impermeability of this wood and its natural resistance to re-wetting.

181

TABLE A.2 Properties of principal construction woods used in northern and southern hemisphere temperate zones

Species	Natural durability Heartwood	Movement classification Heartwood	Treatability or permeability	
			Heartwood	Sapwood
Softwoods				
Abies grandis, Grand fir	ND	S	R	
Agathis robusta, Queensland kauri	ND			
Araucaria angustifolia, Parana pine	ND	M	MR	P
Dacrydium cupressinum, Rimu	MD		R	
Larix decidua, European larch	MD	S	R	MR
Picea spp, European or Canadian spruce, European whitewood	ND	S	R	R
Picea sitchensis, Sitka spruce (fast-grown)	ND	S	R	MR
Pinus spp, Southern or pitch pine	ND	M	MR	P
Pinus banksiana, Jack pine	ND		MR	
Pinus contorta, Lodgepole or contorta pine	ND	S	R	
Pinus lambertiana, Sugar or soft pine		S		
Pinus monticola, Western white pine	ND		MR	
Pinus nigra, Corsican pine	ND	S	P	
Pinus patula, South African pine	ND		P	
Pinus pinaster, Maritime pine	MD		R	
Pinus ponderosa, Ponderosa pine	ND		MR	
Pinus radiata, Radiata or Monterey pine	ND	M	MR	
Pinus resinosa, Red pine	ND		MR	
Pinus strobus, Eastern white pine	ND	S	MR	
Pinus Sylvestris, Scots pine, European redwood slow-grown (northern)	MD	M	R	P
fast-grown (British Isles)	ND	M	MR	P
Pseudotsuga menziesii, Douglas fir or Columbian pine slow-grown (Canadian)	MD	S	R	
fast-grown (British Isles)	ND	S	R	MR
Sequoia sempervirens, Redwood (American)	D		MR	
Taxodium distichum, Southern cypress	D		MR	
Thuja plicata, Western red cedar	D	S	R	R
Tsuga heterophylla, Western hemlock	ND	S	R	MR
Hardwoods				
Acer spp, Maple, sycamore	ND	M	P	P
Afzelia spp, Atzelia	VD(T)	S	ER	
Albizia spp, West African albizia	VD(T)		ER	
Aucoumea klaineana, Gaboon	ND			
Baikiaea plurijuga, Rhodesian teak	VD(T)	S		
Betula pendula, European birch	ND	L	P	P
Carya spp, Hickory	ND			
Castanea sativa, European (sweet) chestnut	D	S	ER	
Chlorophora excelsa, Iroko	VD(M,T)	S	ER	P
Dipterocarpus spp, Gurjun, Keruing	MD	L/M	R/MR	MR/P
Entandrophragma angolense, Gedu nohor	ND/D	S	ER	
Entandrophragma cylindricum, Sapele	MD	M		
Entandrophragma utile, Utile	D	M	ER	
Eucalyptus citriodora, E. maculata, Spotted gum	MD(T)		ER	P
Eucalyptus diversicolor, Karri	MD/D	L	ER	P

TABLE A.2 (continued)

Species	Natural durability Heartwood	Movement classification Heartwood	Treatability or permeability	
			Heartwood	Sapwood
Eucalyptus grandis, E. saligna, Saligna gum	MD		ER	P
Eucalyptus marginata, Jarrah	D/VD(T)	M	ER	P
Fagus sylvatica, European beech	ND	L	P	P
Fraxinus excelsior, European ash	ND	M	MR	P
Gonystylus macrophyllum, Ramin	ND	L		
Gossweilerodendron balsamiferum, Agba	D(M)	S	R	
Guarea cedrata, G. thompsonii, Guarea	D/VD	S	ER	
Juglans regia, European walnut	MD	M	R	
Khaya anthotheca, K. ivorensis, K. nyasica, African mahogany	MD	S	ER	
Nauclea diderrichii, Opepe	VD	S	MR	P
Nesogordonia papaverifera, Danta	MD	M	MR	
Ocotea rodiaei, Greenheart	VD(M,T)	M	ER	
Pericopsis elata, Afrormosia	VD	S	ER	
Quercus spp, American red oak	ND	M	R	
Quercus spp, American white oak	D	M	ER	
Quercus spp, Japanese oak	ND	M		
Quercus cerris, Turkey oak	MD	L	ER	
Quercus robut, Q. petraea, European oak	D	M	ER	P
Shorea spp, dark red meranti or seraya	MD		ER	
Shorea spp, light red meranti or seraya	D	S	ER	
Shorea spp, white meranti			MR	
Shorea spp, yellow meranti or seraya	MD	M	ER	
Swietenia macrophylla, American mahogany	D(T)	S	ER	
Tectona grandis, Teak	VD(T)	S	ER	
Terminalia ivorensis, Idigbo	D	S	ER	
Terminalia superba, Afara, Limba	ND	S	MR	
Tieghemella africana, T. heckelii, Makore	VD	S	ER	
Triplochiton scleroxylon, Obeche	ND	S	R	
Ulmus spp, Elm	ND	M	MR	

Notes

1. *Durability*

VD: very durable	– more than 25 years	(30)
D: durable	– 15–25 years	(31)
MD: moderately durable	– 10–15 years	(32)
ND: non-durable or perishable	– less than 10 years	(33)

Durability refers only to resistance to fungal decay in ground contact, the life in years being based upon field trials in temperate conditions, as detailed in Princes Risborough Laboratory Technical Note No. 40. Durability is quoted only for heartwood; sapwood is invariably less durable. Woods known to possess good resistance to marine borers and termites are marked (M) and (T) respectively.

2. *Movement*

L: large	– more than 4.5%	(27)
M: medium	– 3.0–4.5%	(28)
S: small	– less than 3.0%	(29)

Movement refers to the sum of the percentage changes in dimension in the tangential and radial directions for a change in atmospheric relative humidity from 90% to 60%, as discussed in Princes Risborough Laboratory Technical Note No. 38. The assessment was made on seasoned wood; considerable shrinkage occurs on initial drying for most woods, even those showing only small movement when seasoned. Movement is quoted only for heartwood; sapwood is invariably less stable.

TABLE A.2 (continued)

3. *Treatability*

    ER: extremely resistant   (34)
      R: resistant   (35)
   MR: moderately resistant   (36)
      R: permeable   (37)

Heartwood and sapwood assessments are given where available but, if heartwood alone is listed, sapwood is usually more permeable. The assessment system was generally as in FPRL (Princes Risborough Laboratory) Bulletin No. 54.

4. *Other woods*

A comprehensive list of woods and their properties is given in Appendix 3 of the book *Wood in Construction* by the present author. In this appendix woods are listed according to their Latin names, which can be readily identified by consulting the index to common names. The appendix lists alternative common names, sources and colour, in addition to summarizing typical properties by means of a numerical notation which is suitable to use with a punched card sorting system. For easy cross-reference the appropriate numbers are shown in brackets in the notes above, so that entry in Appendix 3 with 32, 27 and 37 signifies a wood with heartwood that is moderately durable, that has large movement and that is permeable to treatment.

Generally, wood that is fast grown is less durable, less stable and more permeable than slow-grown wood of the same type, as, for example, Scots pine produced in the British Isles in comparison with that produced in North Scandinavia.

Round or sawn

Round poles typically involve relatively impermeable heartwood, perhaps with some natural durability, surrounded by relatively permeable sapwood, which enables a deep preserved zone to be achieved. In sawn wood the surface possesses a random distribution of sapwood and heartwood and, whilst the heartwood may be moderately durable as in slow-grown European redwood, it cannot be readily treated and deterioration will occur when it is exposed to a continuous decay risk, as in ground contact. Thus piles, transmission poles and posts in ground contact should always be made of round wood with a generous zone of permeable sapwood, unless it is possible to use one of the previously-mentioned completely permeable species, such as Southern pine.

Wood-borers

A person involved in the wood preservation industry is normally well acquainted with the wood-borers of local significance but may know little of rare species, particularly if they are found in wood imported from another country. If preservatives or preserved wood are to be exported the habits of particular borers in the destination country may also have considerable importance. The purpose of this appendix is therefore to provide a reasonably thorough description of the borers that are of significance to the European wood preservation industry. It does not, however, include a description of the many other insects which are found, particularly in domestic premises, and confused with wood-borers; a description of these insects appears in *Remedial Treatments in Buildings* by the present author, a book which also includes a more detailed account of the methods for identifying borer damage.

Classification

Before describing the individual borers it is necessary to explain the method of classification that is used by zoologists. The animal kingdom is divided into three sub-kingdoms, the protozoa, the parazoa and the metazoa, but wood-borers are confined to only two of the twenty phyla that comprise the metazoa. The most important phylum is the Arthropoda which includes the most important group, the terrestial insect borers, as well as the marine crustacean borers. The phylum Mollusca includes only a few marine borers. In fact, with the exception of the marine crustacean and molluscan borers, all other wood-borers are terrestrial insects.

Insects – beetles

The class insecta includes five orders containing wood-borers. All insects possess six legs and one or two pairs of wings. The life cycle commences with an egg which hatches to produce a larva which feeds and grows, eventually pupating and metamorphosing into an adult insect. The most important order is the Coleoptera, the beetles. In this order the fore-wings are modified to form hard elytra or cases over the folded hind-wings. The beetles include the most important borers and these can be classified into ten families. The Platypodidae represent one of the two major groups of pinhole borers, the other being the Scolytidae. The Bostrychidae include two very important sub-families, the Lyctidae or Powder Post beetles and the Anobiidae or Furniture beetles. The Cerambycidae or Longhorn beetles are also very important in two ways; many species cause damage to freshly felled green wood and sickly trees in the forest but a single species, *Hylotrupes bajulus*, represents one of the most serious wood-borer pests, in that it infests dry softwood in the warmer temperate zones. The other six families are far less important; they are the Curculionidae or weevils, the Buprestidae or flat-headed borers, the Oedemeridae, which include the wharf borer, and the minor Lymexylonidae, Dermestidae and Tenebrionidae families.

Termites

The termites, order Isoptera, are the most serious wood-borer pests in all tropical and sub-tropical zones. These are all social species living in large communities which include both the winged and the non-winged forms. Although they are known as white ants they are not directly related to the true ants, which are in the order Hymenoptera. The most important Isoptera families are the Kalotermitidae or Powder Post termites, the Rhinotermitidae or Moist Wood termites and the Termitidae, which include both subterranean and mound species. The Hodotermitidae are a semi-desert species known as harvester or forager ants, whilst the Termopsidae are of minor importance and the Mastotermitidae consist of only one species in tropical Australia.

Wasps, bees, ants – moths – flies

The order Hymenoptera, the wasps, bees and ants, includes the Wood wasps, Carpenter bees and Carpenter ants, which bore into wood, as well as various predators which are found in borer galleries. The order Lepidoptera consists of the butterflies and moths, whose larvae, or caterpillars, sometimes bore into wood, usually when it is softened by decay, in order to obtain protection during pupation. The Carpenter moths, however, are true wood-borers as the larvae hatch from eggs laid on the bark, producing galleries in the wood, in which they feed and develop until they pupate and emerge as adult moths. The order Diptera, the true flies possessing only a single pair of fore-wings, the hind-wings being represented by halteres or lumps on the thorax, contains only a few minute species which can be classified as wood-borers, these attacking the cambium of some trees when the bark is damaged.

Marine borers

The class Crustacea includes a number of wood-borers in the order Peracarida. They are all marine borers, similar in appearance to wood lice or fleas. The sub-order Isopoda are the gribbles, which are true wood-borers whereas the sub-order Amphipoda do not damage wood but are often found in gribble burrows.

Within the phylum Mollusca the wood-borers are confined to the class Lamellibranchia-ta, molluscs with a symmetrical body which is normally enclosed by a shell that develops in two parts or valves. In the family Teredinidae, which includes the shipworms and pileworms, these valves or shells are used as cutters to form a burrow, and the body grows progressively to fill the burrow as it is formed. The family Pholadidae, the boring mussels, tend to form rather more shallow burrows than those of the Teredinidae, usually attacking sedimentary rocks rather than wood.

Although there are a great number of species which can be classified as wood-borers there are only a few of serious economic importance, yet it is necessary to consider the minor species as they are sure to create interest and perhaps problems when encountered, and their identification is naturally more difficult than that of the well-known species. Latin names are deliberately quoted as these will enable readers to search for further information if they wish; common names frequently vary in different countries. It must, however, be appreciated that this does not attempt to be a comprehensive account of all wood-borers. Only the borers of significance in wood preservation are described in detail and others are mentioned only in passing so that the reader is aware of their existence.

Ambrosia beetles

According to the systematic approach briefly outlined earlier in this appendix the first wood-borers that must be considered are the Ambrosia beetles, members of the Scolytidae and Platypodidae, which are also known as pinhole or shothole borers after the damage that they cause. The adult beetles are generally 3–6 mm (1/8 in–1/4 in) long and bore round holes through the bark of fresh, green logs, particularly in hardwoods rather than softwoods. These holes sometimes penetrate deep into the wood, forming extensive branched galleries (Fig. B.1), the shape often being characteristic of the species. Eggs are laid in the galleries and the 'ambrosia'

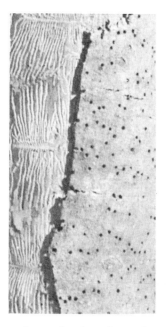

FIGURE B.1 Ambrosia beetle galleries beneath bark. (Wykamol Ltd)

fungus introduced through spores carried by the adult. The development of this fungus is, of course, dependent upon the presence of sugar, starch and a high moisture content in freshly felled wood and this is, in fact, a condition for infestation by these beetles as the larvae hatching from the eggs browse on the fungus. Some larvae extend the galleries to form niches in which they pupate.

Pinhole, Shothole

The damage consists of galleries which are sometimes just beneath the bark and thus commercially insignificant but in some species the galleries penetrate into the sapwood, particularly in tropical hardwoods and, for example, in European oak attacked by *Platypus cylindrus*. The bore holes vary in size, the smaller ones being described as pinholes and the larger ones in tropical species being sometimes known as shotholes. The galleries are free from bore dust and stained brown or black internally by the Ambrosia fungus, and this infection can also extend along that grain adjacent to the galleries to form a characteristic candle-shaped stain. Pinhole or shothole damage is never structurally significant but frequently occurs in decorative woods and results in lower grading.

Infestation ceases as the wood dries and damage can be avoided only by treatment of the logs immediately after felling. Some species of Ambrosia beetles are dependent upon bark and can be readily controlled by its removal. Damage by other species can be prevented to some extent by rapidly drying logs or immersing them in water, or, in the case of temperate climates, by extraction during the winter months. These precautions are not always possible and treatment of freshly felled logs with suitable insecticides is frequently employed, perhaps coupled with a fungicidal treatment to prevent the development of Ambrosia and other staining fungi. A few Platypodids have been found boring in living trees. This rarely happens but some damage to tropical hardwoods may be caused in this way and cannot therefore be controlled by log treatments. Only a few holes may be present in the bark of an apparently sound tree so that the damage cannot be detected until after felling and conversion of the wood.

Oak and chestnut from North America is often damaged and graded as 'sound wormy', whereas damaged mahogany from Africa is graded as 'pin wormy', both gradings indicating that the wood is structurally sound but unsuitable for use in solid furniture or veneer. Damage is normally caused by Scolytids, but some Platypodids are also important, and damage is most severe in sub-tropical and tropical climates. The most important Scolytid genera are *Trypodendron*, *Pterocyclon*, *Webbia*, *Anisandrus* and, particularly *Xyleborus* (Fig. B.2). The most important Platypodid genera are *Platypus*, *Crossotarsus* and *Diapus*. Seventy Scolytid species have now been identified in the British Isles alone, but the most serious damage in Europe can be attributed to a single

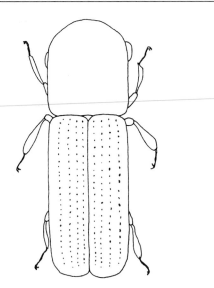

FIGURE B.2 A Scolytid, *Xyleborus saxesini*.

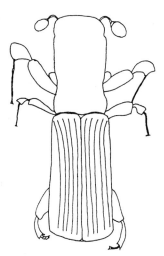

FIGURE B.3 A Platypodid, *Platypus cylindrus*.

Platypodid, *Platypus cylindrus* (Fig. B.3), which is found particularly in more southerly, oak woodlands. This species, which can often be found in wind-blown trees, stumps and logs, principally infests oak but also occasionally beech, ash and elm. *Graphium* spp of fungi are always found in the galleries and often *Cephalosporium* and *Ceratostomella* spp in addition, the latter being well known as a cause

of sapstain, even in the absence of Ambrosia beetle attack. Although several other non-indigenous Platypodids are now found in Europe, almost all other damage can be attributed to Scolytids. *Scolytus (Hylesinus) fraxini* frequently causes damage in ash logs, largely in the cambium immediately under the bark, whilst *Scolytus destructor* causes similar damage in elm.

A beetle that has attracted considerable attention in recent years is *Scolytus multistriatus*, as heavy infestations are always found in trees infected by the Dutch Elm disease fungus, *Ceratocystis ulmi*. As spores of this fungus are always found on these beetles they have been frequently described as the cause of the disease, yet Scolytids do not normally attack healthy trees and it is far more likely that the beetles are attracted to a tree which is already infected and which will thus provide a good site for boring and egg laying. This is perhaps confirmed by the observation that this species is found throughout Europe and North America, as well as in Australia, where the disease is unknown. Generally, the species forms a vertical gallery under the bark of a standing tree, with larval galleries branching off to form a characteristic fan shape.

Other Scolytids causing damage to hardwoods in Europe are *Xyloterus signatus* and *domesticus*, *Xyleborus saxeseni* and *monographus*, and *Anisandrus dispar*.

Scolytids also cause damage to softwoods, for example *Ips typographus* and *Pityogenes bidentatus* in spruce and *Myelophilus* spp in pines in the British Isles. Occasionally, severe local infestations have occurred, such as that of *Trypodendron lineatum* in Argyllshire, but in Europe generally *Xyloterus lineatus* probably represents the most serious Ambrosia beetle problem in softwood. In North America the most serious damage is caused by *Dendroctonus* spp, *D. frontalis* being known as the Southern pine beetle, *D. breviconis* as the Western pine beetle and *D. ponderosae* as the Mountain pine beetle respectively; damage by all of these species occasionally being observed in wood imported into Europe. Generally, a wide variety of species is involved in damage caused in imported tropical hardwood, but occasionally an individual species may attract particular interest, such as *Diapus furtivus*, which was at one time a particular problem in hardwood imported from Malaysia.

Melandryid bark-borers

The family Melandryidae are bark-borers, and are often confused with the Ambrosia beetles or the Anobid bark-borer, *Ernobius mollis*, which will be described later. Thirteen genera of very variable form and habits have been reported in the British Isles but they are all rare and the only species likely to be observed as a bark-borer is *Serropalpus barbatus*.

Bostrychid family

The family Bostrychidae includes a number of important species which can be most conveniently divided into two groups, the Powder Post beetles and the Furniture beetles. The Powder Post beetles can be divided in turn into two sub-families, the Bostrychids and the Lyctids.

Bostrychid Powder Post beetles

The Bostrychids, the Auger beetles, are also known as Shothole borers in some areas such as South Africa and this can lead to confusion with the Ambrosia beetles. However, in the areas concerned the Ambrosia beetles, the Scolytids and Platypodids, are invariably small and known as Pinhole borers so that the possible confusion is confined only to the larger borers. Most Auger beetles are small, 3–6 mm (1/8 in–1/4 in) long, although one species, *Bostrychopsis jesuita*, is about 20 mm (3/4 in) long. The beetles are cylindrical in section with spines on the front edge of the thorax, which is hooded so that is conceals the head from above, the bostrychoid form, from which this sub-family and family derive their name. These features and the three-jointed

antennae enable the Bostrychids to be readily identified. The European Bostrychids are dark brown or black in colour with the single exception of a species attacking oak, *Apate capucina*, which has brown or red elytra.

Adult Bostrychids tunnel in bark in order to lay eggs, producing tunnels which are free of dust. The hatching larvae then bore in the sapwood in search of starch, producing tunnels which are packed with fine bore dust, as is the case with the Lyctid Powder Post beetles. This pattern of tunnelling and the four-jointed legs of the curved larvae enable this damage to be distinguished from that of the Lyctids. In fact, Bostrychid damage is not so common as Lyctid damage, probably because infestation commences with a tunnel bored by the adult in contrast with a Lyctus infestation which is initially completely invisible. Damage is principally confined to the sapwood of green hardwoods, although softwoods are occasionally found to be attacked, particularly if they have bark adhering. The adult beetles are able to bore into wood treated with some preservatives, such as creosote and copper–chromium–arsenic salts, but the egg larvae die and are unable to cause further damage.

The most common Bostrychid damage in tropical hardwoods is caused by *Heterobostrychus brunneus*, *Xylopertha crinitarsis* and *Bostrychoplites cornutus* in West African woods and *Heterobostrychus aequalis* in woods from India, Malaysia and the Philippines. North American hardwoods are infested by *Schistoceros hamatus* and occasionally *Xylobiops basilare* in ash, hickory and persimmon. The only Bostrychid common in Europe is *Apate capucina* (Fig. B.4), which attacks oak. *Dinoderus* spp, particularly *D. minutus*, are sometimes found in bamboo and basketwork. In Australia *Mesoxylion collaris*, a red-brown Bostrychid about 6 mm (1/4 in) long is sometimes found in the south-east, attacking the sapwood of sawn wood in buildings, and *Mesoxylion cylindricus*, a larger and deeper

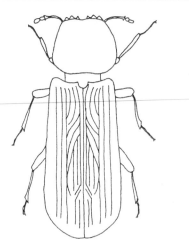

FIGURE B.4 A Bostrychid, *Apate capucina*.

brown Bostrychid about 12 mm (1/2 in) long is sometimes found in poles and mining timber. The very large *Bostrychopsis jesuita*, a dark brown or black beetle 20 mm (3/4 in) long and producing 6 mm (1/4 in) diameter holes is often found in Australia. It is not common in sawn wood but is occasionally found in sub-floor structures, where the relatively high moisture content permits its continuing development.

Lyctid Powder Post beetles

The Lyctid beetles are all small with a length of only about 4 mm (1/6 in). These beetles are elongate but flattened in appearance and, from above, the head is clearly apparent, protruding in front of the thorax. The colour for the various species varies from mid-brown to black and the antennae possess a distinct two-jointed club. The beetles are active fliers, particularly on warm nights. After mating the female lays 30–50 eggs in large pores on the end-grain of suitable hardwoods containing adequate starch. The eggs are elongate, oval or cylindrical with a strand at one end and they hatch after 8–14 days. The larvae feed on the yolk at first until they grow sufficiently to gain a purchase on the side of the pore. At this stage the larva is straight-bodied and moves along the grain but, after an initial moult, it becomes curved and

commences to burrow across the grain, and eventually grows to a length of about 6 mm (1/4 in). It can be distinguished from other similar larvae by the prominent spiracle on either side of the eighth segment, immediately before the last segment, and the three pairs of minute three-jointed legs.

Pupation occurs in a chamber immediately beneath the surface and an adult beetle emerges after 1 or 2 years in the spring, summer or autumn, usually between late May and early September in Europe, leaving a flight hole 0.8–1.5 mm (1/32 in–1/16 in) in diameter. The life cycle is shorter in warmer conditions. The galleries are packed with soft, fine bore-dust but the tunnels are not distinctly separate as are those of the Furniture beetles and all the sapwood may be completely destroyed except for a surface veneer, accounting for the name of Powder Post beetle. As the initial attack consists only of an egg laid in an open pore it will be appreciated that the first sign of damage is collapse or alternatively the appearance of a flight hole, which is an indication of extensive damage within the wood.

It is unlikely that Lyctid beetles are indigenous in Europe and new species are being continually introduced in infested wood. The two most important species are *Lyctus brunneus* (Fig. B.5) and *L. linearis*, the latter being readily distinguishable by the long rows of hairs on the elytra. *Lyctus planicollis* and *L. parallelopipedus* were originally confined to North America but have been identified extensively in Europe since World War I, whilst *L. cavicollis* from the United States and *L. sinensis* from Japan have also now been introduced in imported oak. *Minthea* spp, which are largely confined to tropical hardwoods, are difficult to distinguish from *Lyctus* spp and cause exactly the same damage; *M. rugicollis* is most common. However, *Minthea* adult beetles burrow, as does *Lyctus africanus*, whereas only the larvae of other Lyctids do so.

The Bostrychid Powder Post beetles will at-

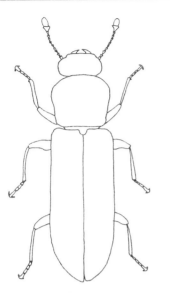

FIGURE B.5 A Lyctid, *Lyctus brunneus.*

tack most freshly felled hardwoods, preferably those with bark adhering, provided the moisture content remains reasonably high and stable. In contrast, the Lyctid Powder Post beetles confine their infestations to dry, felled wood still retaining starch but only of those species which have large pores, such as oak, ash, walnut, elm and hickory in temperate climates. Numerous tropical woods are susceptible, particularly light-coloured species. Susceptible African species are obeche (wawa), agba, afara (limba), African mahogany, antiaris, iroko, afzelia, albizia, panga panga, gaboon (okoume), Rhodesian teak and, in South Africa, black wattle. In the Far East susceptible woods include ramin and the many *Shorea* spp, such as seraya, meranti and lauan. In Australia many of the Eucalypts are susceptible.

Starch is essential to the development of all Powder Post beetles, whether Lyctids or Bostrychids. In temperate forests the felling of hardwoods in early winter when the starch content is higher will obviously encourage infestation. Starch is confined to the sapwood so that sapwood removal represents one method of control, but usually insecticide treatment is applied if wood is to be air-seasoned. The starch degenerates progressively and wood becomes immune to attack after a period of several years. Air-seasoned wood is, in fact, less susceptible as the cells remain alive during the early stages of seasoning, reducing the starch reserves. Kiln-seasoning is sometimes advocated as a means of controlling Powder Post beetles but, whilst it achieves complete control of Bostrychids, the wood cells are kilned at temperatures in excess of 40°C (104°F) allowing the starch to remain so that kiln-dried wood is particularly susceptible to a new infestation.

Heavy Powder Post beetle infestations are generally accompanied by parasites and predators, including the mite, *Pyemotes (Pediculoides) ventricosus*, the Clerid beetles, *Tarsostenus univitartus* and *Paratillus carus*, as well as ant-like Hymenoptera, the wingless *Sclerodermus domesticus* and *S. macrogaster*, and the winged *Eubadizon pallipes*; predators are mentioned later in this appendix.

Anobid Furniture beetles

The Furniture beetles, comprising the sub-family Anobiidae, are probably the best known woodborers in temperate areas, probably because damage occurs in furnishings and is thus readily apparent to the householder. This sub-family can be further divided into the Anobiinae, which produce elongated ovoid or rod-shaped pellets, and the Ernobiinae, consisting of *Ernobius, Xestobium* and *Ochina* spp, which produce bun-shaped pellets, a useful diagnostic feature when only damaged wood is available. All the Furniture beetles possess the hooded bostrychoid thorax concealing the head from above. The last three joints of the antennae are always larger than the other joints, except in *Ptilinus pectinicornis* and *Lasioderma serricornia*, which are readily identified by their comb-shaped antennae. All the larvae are curved and the life cycles tend to be long, with a very slow build-up of infestation over many years. All Furniture beetles lay eggs in cracks or open pores and consid-

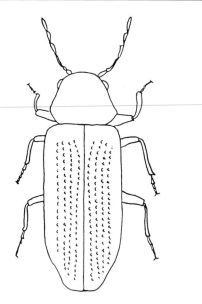

FIGURE B.6 Common Furniture beetle, *Anobium punctatum.*

erable damage may occur before the infestation becomes apparent through the development of flight holes and bore dust discharges.

Anobium punctatum (domesticum, striatum), the Common Furniture beetle (Fig. B.6), probably originates in the northern temperate zones but it is now widely distributed throughout the world, infesting sapwood of hardwoods and softwoods, as well as heartwoods of some light-coloured hardwoods and heartwood of most species in damp conditions. The beetle is 2.5–5.0 mm (1/10 in–1/5 in) long – the female tending to be larger – reddish to blackish brown in colour and with a fine cover of short, yellow hairs over the thorax and elytra, particularly on freshly emerged specimens. When viewed from the side the hooded thorax has a distinct hump and there are also rows of pits on the elytra which account for the names *punctatum* and *striatum.*

The beetles emerge from the wood in northern temperate climates between late May and early August and can often be seen crawling on walls and windows. The beetles are strong fliers on warm days and live for 3–4 weeks during which they mate and the female lays up to 80 lemon-shaped, white eggs about 0.3 mm (1/80 in) long in cracks, crevices, open joints and old flight holes, usually in small groups. The eggs hatch after 4–5 weeks, the larva breaking through the base of the egg and then tunnelling within the wood in the direction of the grain. The gallery increases in diameter as the larva grows, occasionally running across the grain. The galleries are filled with loosely-packed, gritty bore dust consisting of granular debris plus oval or cylindrical pellets, compared with the fine powder in the case of Powder Post beetle attack. When fully grown the curved larva is about 6 mm (1/4 in) long with five-jointed legs. Eventually the larva forms a pupal chamber near the surface about 6–8 weeks before emergence through a flight hole about 1.5 mm (1/16 in) in diameter, larger than a Lyctid flight hole but smaller than that of the Death Watch beetle. Under optimum conditions the life cycle can be as short as one year but it is usually longer and up to four years.

As infestation is largely confined to sapwood the damage is not usually structurally important, except where individual components in furniture are composed entirely of sapwood or where old types of blood or casein adhesives have been used, as these considerably encourage infestation. Infestation is also encouraged by dampness, slight fungal or bacterial activity enabling the infestation to extend into normally resistant heartwood. All these situations tending to favour Common Furniture beetle attack appear to be related to nitrogen availability and result in shorter life cycles; such an exaggeration of activity is particularly noticeable in stables and byres, which are often extensively damaged. As in the case of other Anobids, activity is indicated only when a bore dust discharge suggests recent emergence. It is therefore often difficult to decide whether a remedial treatment is necessary or, if it has been completed, whether it has been effective.

Common Furniture beetles (Fig. B.7(a)) suf-

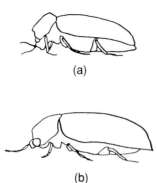

(a)

(b)

FIGURE B.7 Side views of (a) Common Furniture and (b) Death Watch beetles, showing distinctive shapes of the hooded thorax.

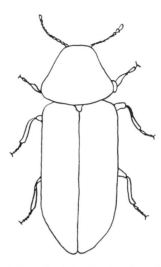

FIGURE B.8 Death Watch beetle, *Xestobium rufovillosum.*

fer a very high natural mortality, perhaps only 60% reaching the larval stage, where they may be further reduced in numbers by the action of predators. The most common predators are two Hymenoptera, the flying, ant-like *Theocolax formiciformis* and *Spathius exarator*, which can often be seen exploring flight holes. Predatory beetles are also sometimes found, such as *Opilio mollis (domesticus)* and *Korynetes coeruleus*, but the latter is more often associated with the Death Watch beetle; predators are described later in this appendix.

In New Zealand *Anobium punctatum* is known as the Common Houseborer, an unfortunate name as it leads to confusion with the House Longhorn beetle, which is known in Northern Europe and many other countries, such as South Africa, as the Houseborer. The Common Furniture beetle is a particularly serious problem in New Zealand as it can cause extensive structural damage to the light-framed, softwood buildings that are so often used. The closely related *Anobium pertinax* is sometimes found, particularly in buildings in Scandinavia, but only in association with fungal attack so that severe infestations are normally confined to poorly ventilated cellars or sub-floor spaces, or in timber subject to periodic rainwater or plumbing leaks.

Xestobium rufovillosum (tesselatum), the

Death Watch beetle, (Figs B.7(b) and B.8), is the largest Anobid, with a length of 6–8 mm (1/4 in–1/3 in). The Death Watch beetle attacks only wood that is subject to dampness and some decay; indeed, a common characteristic feature of Death Watch beetle attack is a brown colouration in the infested wood, arising from the fungal decay. Infestations in the British Isles occur most commonly in oak, probably because this wood used to be extensively employed in construction, but infestations can also occur in elm, walnut, chestnut, alder and beech. Sapwood and heartwood can be infested if previously infected by decay, and the infestation can spread into adjacent softwoods, though infestation is always confined to damp or decayed areas. It frequently appears that this insect favours churches but this is really a combination of circumstances which results in church timbers being particularly suitable; the roofing frequently consists of sheets of lead or other metals and the periodic heating results in condensation, which causes the incipient decay that encourages infestation.

The Death Watch beetle is chocolate brown in colour and has patches of short, yellow hairs, which give a mottled appearance. The thorax conceals the head but is very broad, appearing from above to form a hood over the front ends of the elytra, whereas other Anobids have a distinct waist between the thorax and the elytra. The larvae are very similar to those of the Common Furniture beetle but attain a far larger size, eventually growing to about 8 mm (1/3 in).

The length of the life cycle depends upon the quantity of nitrogen available in the form of fungal attack and can be a single year under optimum conditions, but it is usually far longer and perhaps as much as ten years. In northern temperate climates the adult emerges between the end of March and the beginning of June and, after mating, the female lays about 40–60 white, lemon-shaped eggs 0.6 mm (1/40 in) long in cracks, crevices and old exit holes. The larvae hatch after 2–8 weeks and explore the surface of the wood before commencing to bore. Where longer life cycles are involved the fully-grown larva pupates in July or August, metamorphosing into an adult after only 2–3 weeks, but it remains in the wood and gradually darkens in colour until it emerges the following spring, leaving a flight hole 3 mm (1/8 in) in diameter.

The adult is not a free flier and tends to mate with other beetles emerging from the same piece of wood, attracting their attention by moving the legs so that the head is struck on the wood surface, producing a series of 8–11 taps in a period of about 2 seconds. Tapping with the tip of a pencil can generate a similar noise and can stimulate a response. This tapping noise probably accounts for the name of Death Watch beetle, perhaps because the sound is apparent in a house which is quiet through a recent death. The tapping should not be confused with the sound produced by Psocids such as the book lice, *Trogium pulsatorium*, which is more like a watch tick than a tap. Although the insects can fly reasonably well when the weather is very warm, it is probable that this pest is spread largely by re-use of old infested wood.

The Death Watch beetle has a much more

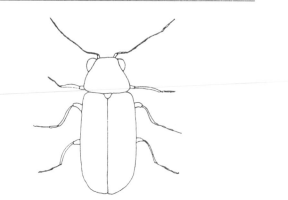

FIGURE B.9 The Barkborer, *Ernobius mollis*.

limited distribution than the Common Furniture beetle and in the Britih Isles it is confined to England, Wales and part of southern Ireland. In some areas, such as Germany, the Death Watch beetle is frequently confused with the Common Furniture beetle although the flight holes are much larger and the galleries are packed with much coarser bore dust with distinct bun-shaped pellets, as opposed to the oval or cylindrical pellets of the Common Furniture beetle. The Death Watch beetle is always associated with dampness and decay and can therefore be readily distinguished from *Ernobius mollis*, which produces a similar sized flight hole, as the latter is confined to softwoods with adhering bark. Wood infested by the Death Watch beetle may also be attacked by other insects, such as the Common Furniture beetle and the Wood weevils, and perhaps *Helops coeruleus* in the damper zones of the wood. Predators may also be present, particularly *Korynetes coeruleus*, a distinctive blue beetle which is an active flier and often the first sign that a Death Watch beetle infestation is present in concealed damp timbers.

Ernobius mollis, sometimes known simply as the Barkborer (Fig. B.9), is an Anobid beetle 3–6 mm (1/8 in–1/4 in) long, midway in size between the Common Furniture and the Death Watch beetles. It is reddish brown in colour with yellow hairs. Compared with that of the Common Furniture beetle the thorax forms a distinct triangle and is not so hooded so that the head protrudes slightly when seen from above. The antennae are also proportionately longer. The larvae are similar in appearance and eventually grow to about 6 mm (1/4 in) long.

Eggs are laid in cracks in the bark of freshly felled softwood, the larvae tunnelling through the bark and up to about 1 cm (1/2 in) into the sapwoods. This is typical of European observations but this beetle also occurs in South Africa, Australia and New Zealand, where it is reported to bore rather more deeply, perhaps because the softwood species in these countries have a deeper sapwood zone. The galleries are filled with bore dust which contains bun-shaped pellets which are either brown or white in colour, depending upon whether the larva has been feeding on bark or xylem. The life cycle is typically one year, the adult beetle emerging in northern temperate climates between May and August from a flight hole 2.5 mm (1/10 in) in diameter usually close to an area of retained bark. This beetle is widely distributed throughout Europe but attacks only softwood with bark adhering to it, often causing the bark to strip away from rustic poles and posts. It is dependent upon starch and, just as the Lyctid Powder Post beetles, can attack wood only for a few years after it has been felled. In remedial treatments damage by this beetle is often confused with that by the Common Furniture beetle.

The Anobid beetle *Ptilinus pectinicornis* (Fig. B.10) is readily identified as it has a distinctly globular thorax and the antennae are comb-like in the male and serrated or saw-like in the female. The adult beetles are distinctly cylindrical, about 3–5 mm (1/8 in–1/5 in) long, with a dark brown prothorax and reddish elytra. The larvae are similar to those of the Common Furniture beetle, but identification can be achieved if it is essential. In many respects the behaviour of this insect is similar to that of the Lyctid and Bostrychid Powder Post beetles. Adult beetles emerge from the wood in Europe between May

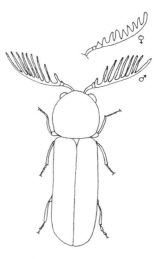

FIGURE B.10 *Ptilinus pectinicornis.*

and July through exit holes about 1.5 mm (1/16 in) in diameter, the same as Common Furniture beetles. The adult female beetles are often found within the wood, apparently extending old galleries or excavating new ones from which to lay their long, thin, pointed eggs in adjacent vessels or pores. The life cycle can be only one year in optimum conditions but it is usually 3–4 years.

This beetle is found in beech, sycamore, maple and elm and can be a nuisance when it occurs in furniture. It is rarely observed in softwoods, clearly because of the limited size of the pores, but it appears that it may occasionally lay eggs in splits. The bore dust is finer than that of the other Anobid beetles and similar to that produced by the Lyctid beetles but more densely packed. This beetle is frequently found in association with the Common Furniture beetle when infesting woods attractive to both species, such as beech.

There are a number of other Anobid beetles which are found in nature but rarely in structural or decorative wood, for example *Nicobium castaneum*, which is found infesting wood in Mediterranean countries, *Anobium denticolle* and *Hedobia imperialis*, which are found in hawthorn, and *Ochina hederae (ptinoides)*, which is found in ivy. *Stegobium panaceum (Sitrodrepa panicea)*, the Drug Store, Biscuit or Bread beetle, is found in dried, woody, natural drugs, cork, dog biscuits and other similar materials, often appearing in large numbers in buildings and producing fears of an extensive Common Furniture beetle infestation; the beetle is, in fact, less elongate, more reddish brown and smaller, and lacks the distinct hump which is a feature of the thorax of the Common Furniture beetle. Similar fears may arise in warehouses where stored tobacco is infested by the Cigarette beetle, *Lasioderma serricornia*, although this Anobid beetle is readily distinguished by its serrated antennae. The closely related Ptinids or Spider beetles are also sometimes confused with Anobids, although they possess an abdomen which is distinctly more rounded, and much longer legs; household pests that are confused with wood-borers are described more fully in *Remedial Treatments in Buildings* by the present author.

Cerambycid Longhorn beetles

The Cerambycidae, the Longhorn beetles, are a very large family with widely varying habits. Their name arises from the fact that their antennae are sometimes longer than the rest of the body. These beetles vary in length from 6 to 75 mm (1/4 in to 3 in) and some species are brightly coloured. The eggs are white, oval or spindle-shaped, and normally laid in crevices in bark. All the larvae are straight, with a slight taper towards the rear. Larvae can be up to 100 mm (4 in) long when fully grown, legless or with very short, useless legs. The flight holes are characteristically oval. Although there are many species there are only a limited number of economic significance as they are mainly forest scavengers infesting damp, rotted wood.

Some Longhorn beetles, such as *Callidium*, *Phymatodes*, and *Trinophylum* spp, bore as larvae under the bark and then penetrate up to 100 mm (4 in) into the wood in order to pupate. Others, such as *Ergates*, *Macrotoma*, *Cerambyx*,

Monohammus and *Batocera* spp, bore entirely in wood, often into heartwood in species such as hickory and ash where there is relatively little differentiation between heartwood and sapwood. Only a few species, such as *Hylotrupes*, *Stromatium* and *Oemida* are able to bore in dry wood after removal of the bark. Often the form of the galleries enables the infesting Longhorn to be identified, so that damage by *Hylotrupes*, *Phymatodes* and *Monohammus* can be readily distinguished in this way. However, the thickness of the bark and girth of the tree can influence the shape and nature of galleries and chambers. For example, the larvae of *Leiopus nebulosus* form a pupal cell as an oval excavation immediately beneath the moderately thick bark in oak but penetrate deep into the sapwood in chestnut, which has a thin bark.

When Longhorn damage is discovered it is most important to decide whether the infestation is confined to the forest or whether it is able to progress and spread in wood in service. As wood dries the life cycle of some of the forest Longhorns increases and many species are able to survive, occasionally for exceptionally long periods of 25 years or more, eventually emerging and perhaps establishing a new Longhorn infestation problem in adjacent suitable forests. Most of the serious Longhorn forest pests in Europe have been introduced in this way.

In Europe hardwoods are attacked by a number of Longhorn beetles, but infestations in healthy trees are rare and in most cases the trees are sickly or the wood even dead or decayed. Oak sapwood is sometimes attacked by the Oak Longhorn beetle, *Phymatodes testaceus* (Fig. B.11). Eggs are laid in cracks within the bark and the larvae bore between the bark and the wood in standing, sickly trees, eventually pupating in chambers within the wood. This species is also sometimes found in beech. Oak is also attacked by *Cerambyx cerdo*, *Clytus arcuatus* and *C. arietus*, the latter being known as the Wasp beetle because of its distinctive

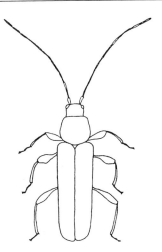

FIGURE B.11 Oak Longhorn beetle, *Phymatodes testaceus.*

black and yellow bands. If oak is decayed it can be attacked by *Rhagium mordax* and *Leiopus nebulosus.*

Sometimes, species associated with softwoods are reported in hardwoods, for example *Hylotrupes bajulus* and *Callidium violaceum* in oak sapwood and *Molorchus minor* under the bark of birch. Aspen, widely used in Scandinavia for the manufacture of matches and sauna furniture, is sometimes attacked by *Clytus rusticus,* whilst *Saperda carcharias* is found in poplar and willow. The latter species is rather unusual as the adults feed on the leaves, tending to kill the tree. The eggs are eventually laid at the base of the young, sickly trees and the larvae bore within the wood, principally along the grain; their galleries are often eventually disclosed by woodpeckers.

The most important Longhorn species to attack North American oak are *Smodicum cucujiforme, Chion cinctus* and *Romaleum rufulum,* these often penetrating the heartwood as well as the sapwood. *Eburia quadrigeminata,* which is yellow or pale brown and 18–24 mm (3/4 in–1 in) long, is also occasionally found and can survive for many years in service before it emerges; it has been found emerging from a bedpost 25 years after manufacture and has been reported as emerging from other items as much as forty years often their manufacture. North American hickory and ash are sometimes infested by the ash borer, *Neoclytus caprea* or the red-headed ash borer, *N. erythrocephalus.*

Longhorn infestations are less common in softwoods, the most serious being that of *Hylotrupes bajulus,* the House Longhorn beetle, which will be considered in detail separately in view of its ability to infest dry softwood in service. Occasionally *Tetropium castaneum* and *T. fuscum* are found in spruce and silver fir from Poland, and *Callidium violaceum* and *Monohammus* spp in pine and spruce from Scandinavia and Russia; such infestations in imported wood have now resulted in these pests becoming established in the British Isles. Other Longhorns introduced in this way are *Tetropium gabrieli,* which is found in England and Wales attacking sickly larch trees and logs which are also attacked by the Spruce Longhorn, *Callidium violaceum. Acanthocinus aedilis* is a cause of damage in pine in Scotland only; this wood is more often attacked by *Criocephalus rusticus, Asemum striatum* and *Rhagium bifasciatum,* the latter being one of the commonest Longhorns in Britain, though it has little economic significance as it attacks only decayed softwoods.

Longhorn infestations are rare in softwood imported from North America, but occasionally Douglas fir or Sitka spruce may be infested by *Monohammus titillator, M. scutellatus* or *Ergates spiculatus;* the latter species is able to survive in wood in service for many years and has been found by the author emerging from Douglas fir floorboards thirty years after installation. Another species that is able to survive in service is the Two-toothed Longhorn from New Zealand, *Ambeodontus tristus,* which is described in more detail later.

Longhorns are not confined to temperate areas. *Oemida gahani* is often found in softwood, such as Cupressus, in Kenya, where *Androeme plagiata,* which is very similar in appearance, is found in both hardwoods and softwoods. By about 1958 Longhorn damage had become so extensive in Kenya in some woods such as *Isoberlinia* that felling had to be abandoned as uneconomic. The Indian Dry-wood borer *Stromatium barbatum* is remarkable as it is known to attack over three hundred wood species. One Indian Longhorn, *Trinophylum cribratum*, was reported to have become established in England in 1947 on hardwoods such as beech; it was apparently introduced on imported wood and causes damage similar in appearance to that by *Phymatodes testaceus* except that the galleries penetrate 50 mm (2 in) or more into the wood and it is therefore of greater economic importance. Longhorn damage is rarely found in tropical African hardwoods and then it is usually caused by *Coptops aedificator* or *Plocaederus* spp, the latter species being characterized by calcareous cocoons which line the pupal chambers in the bark of woods such as mahogany, a feature that is characteristic also of another Longhorn, *Xystrocera*.

It will be appreciated that many of these forest Longhorns are scavengers which infest decaying wood and which are of no significance for wood in service, where the decay problem is far more serious than any new borer infestation. Most of the other species mentioned are bark-borers which attack sickly trees or felled logs and cause little damage unless their pupal chambers penetrate deeply. A few species bore much more deeply in wood but infested wood is usually readily detected and rejected, although occasionally infestations survive in shipments. The principal significance of any such survival is that it can enable infestations to spread to new forest areas; these borers are not generally able to re-infest dry wood. Only a very few species are therefore likely to be encountered as active infestations in wood in service.

Occasionally, active infestations of *Eburia quadrigeminata* and *Ergates spiculatus* may be found in North American oak and Douglas fir, respectively, many years after conversion but

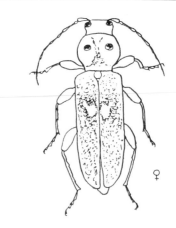

FIGURE B.12 House Longhorn beetle, *Hylotrupes bajulus*.

these are examples of exceptional survival rather than re-infestation. The Oak Longhorn beetle, *Phymatodes testaceus*, is able to attack European hardwoods, particularly oak, after they are dry but only provided the bark remains. The eggs are laid in the bark and the larvae tunnel between the bark and the wood, making deeper tunnels to provide a pupation chamber. This species constitutes a problem principally during air-seasoning, but emergence can also occasionally occur from new oak boards in service; North American Longhorn species may sometimes emerge from imported oak. The two smallest Longhorns, *Cracilia minuta* and *Leptidea brevipennis*, can also attack dry wood, but they are confined to wicker work and are therefore of limited importance. In fact, the only Longhorn species of real economic significance is the House Longhorn beetle (Fig. B.12).

The House Longhorn beetle, *Hylotrupes bajulus (Callidium bajulum)*, is known in many areas as the European Houseborer. Although originally confined to central and southern Europe this species has now been introduced on imported wood to North America, South Africa and Australia, it apparently reaching the latter continent through the importation of infested prefabricated buildings in 1948. Even in Europe its distribution has been influenced by the establishment of trading routes and local conditions. For example, it was first reported in London in 1795, perhaps being imported in orange boxes from Spain, but the infestations apparently died out, probably through pollution in the nineteenth century. It is now confined to the less polluted areas to the south-west of London, where it is known as the Camberley borer, and spread to other parts of England has not occurred, perhaps through climatic restrictions. In Scandinavia it is found throughout Denmark, the Baltic Islands and in the south of Sweden and Norway around the major trading areas such as Oslofjord, Stavanger and Bergen.

House Longhorn beetle attacks the sapwood of dry softwoods. It is not as widespread as the Common Furniture beetle but the damage it causes is much more rapid and more severe so it can be considered the most important wood-borer in temperate areas free from termites. The adult beetle is somewhat flat and (10–20 mm) (2/5 in–4/5 in) long, the male being smaller than the female. The beetles are brown to black in colour except that they have thick, grey hairs on the head and prothorax (the front section of the thorax), and the female has a central black line and a black nodule on either side and the male white marks. There are also distinct, shaped white spots on the elytra.

In Europe the beetles emerge in July to September and a single female can lay as many as 200 eggs, which hatch within 1–3 weeks, these eggs being spindle-shaped and 2 mm (1/12 in) long. In a roof structure the larvae from a single clutch of eggs can cause substantial damage within a period of 3–11 years before they pupate and emerge as adults, perhaps entirely destroying the sapwood and leaving only a thin surface veneer, this slightly distorted by the presence of the oval galleries beneath. The first sign of damage may therefore be the collapse of a largely sapwood member, though in warm weather the gnawing of the insects can be clearly heard. When fully grown the larva is about 30 mm (1¼ in) long, straight-bodied and dis-

tinctly segmented, with a slight taper and very small legs. Pupation occurs in a chamber just below the surface and is complete in three weeks, the emerging beetle leaving an oval flight hole about 1 cm (3/8 in) across.

The appearance of even a single flight hole indicates that severe damage has already occurred and that the condition of the structure should be checked by thorough probing. Because of the seriousness of the damage this beetle causes, the building regulations in England now require all structural wood to be preserved against this pest in the areas to the south-west of London where it is known to occur. Similar regulations have been introduced in other European countries and in South Africa, whilst the Australian quarantine regulations are designed to prevent further infestations being introduced.

Two-toothed Longhorn beetle

The Two-toothed Longhorn, *Ambeodontus tristus*, causes serious damage to softwoods in service in New Zealand. It thrives in similar conditions to the House Longhorn beetle and causes similar damage, but the oval exit holes are distinctly smaller, being only about 5 mm (1/5 in) across. In 1974 this insect was found to have caused severe damage to joists in a cellar in Leicestershire, England. The infestation was introduced in the joists, which were found to be made from the wood of a *Dacrydium* pine, common in New Zealand.

Curculionid weevils

As wood becomes damp or decayed the activity of many wood-borers, such as the Common Furniture beetle, is encouraged and the wood may become infested by other species dependent upon decay, such as the Death Watch beetle and particularly the wood-boring weevils, Curculionidae (Cossonidae). The weevil is a shiny, cylindrical beetle, 3–5 mm (1/8 in–1/5 in) long and brown to black in colour, its head protruding into a long snout with 'elbowed' antennae about halfway along its length. The larvae are curved and legless.

Both the adults and the larvae bore, causing damage which is superficially similar to that made by the Common Furniture beetle but the holes are smaller, contain finer bore dust, are often laminar in pattern and are perhaps confined to the spring wood. The attack is often limited to sapwood but extends into heartwood if fungal decay is more severe; however visible fungal decay does not always appear to be necessary. Eggs are laid singly, either in small holes in the surface or in niches within the galleries, at any time of the year and they are hatched within 2–3 weeks, the larvae then boring for 6–12 months before metamorphosing into adults, usually during June to October. Unlike most wood-boring beetles the adults can survive for a very long period, perhaps 12 months or more, actively tunnelling within the wood.

In Britain *Pentarthrum huttoni* (Fig. B.13) is a native species usually found in buildings in decayed floorboards and panelling, as well as in old casein-glued plywood. *Caulotrupis aeneopiceus*, another native species, is more rarely identified in buildings and is then only found in association with very decayed wet wood in cellars and under-floor spaces. Very rarely infestations may be found to be due to *Cossonus ferrugineus* and *Rhyncolus lignarius*. However, in recent years the weevil that has attracted most attention is *Euophryum confine*. This species was apparently introduced to Britain from New Zealand in about 1935 and it has since spread very widely, apparently because it is able to infest wood which is not significantly decayed and which may have a moisture content as low as 20%. This species is therefore of greater significance. Adult beetles of *Euophryum confine* can be distinguished from those of *Pentarthrum huttoni* by the sharp constriction of the head behind the eyes, as illustrated in Fig. B.13.

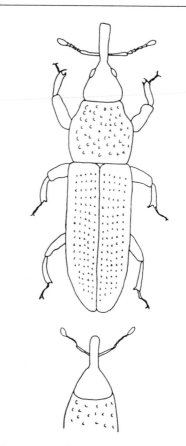

FIGURE B.13 Wood weevils, *Euophryum confine* and the head of *Pentarthrum huttoni*.

Oedermerid beetles

The Wharf borer, *Nacerda melanura* (Fig. B.14), a member of the family Oedermeridae, is superficially similar to a Longhorn beetle. It is a free flier and has sometimes been found in great numbers in streets and buildings close to dock areas in, for example, London and Copenhagen. The beetle attacks both softwoods and hardwoods which are decayed and apparently favours wood which is wetted by sea water, brine or urine, and the large numbers which are sometimes found originate from piles, groins, quays and piers in sea or river areas. One of the mysteries is that, at other times, this insect is rarely observed and it is therefore sometimes

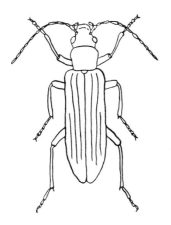

FIGURE B.14 Wharf borer, *Nacerda melanura*.

confused with the carniverous soldier beetles, *Rhagonycha fulva*, which are common in summer on garden flowers.

The Wharf borer is 6–12 mm (1/4 in–1/2 in) long, elongate and reddish brown with distinct black tips to the elytra and long antennae. The soldier beetles are much redder with a wider prothorax, less flattened and without the lateral flanges possessed by the Wharf borer. The Wharf borer larvae are rather slender, eventually achieving a length of 12–30 mm (1/2 in–1¼ in), dirty white in colour, and have a large yellow head and three pairs of fairly long legs. The first segment behind the head has a distinct hump or knob in the centre of the back. Damaged wood is usually brown in colour through decay but also distinctly laminar as most of the attack takes place along the grain.

Other beetles

Another insect associated with damp, decayed wood is *Helops coeruleus*, one of the Tenebrionidae. Attack by this insect is rare, at least in the British Isles, where it is confined to a few areas in southern and eastern England. In structural woodwork it is almost always associated with the Death Watch beetle in oak or chestnut, the Death Watch beetle usually attacking areas which are reasonably sound but

infected with fungi such as *Merulius lacrymans*, *Coniophora cerebella* and *Phellinus megaloporous*, whereas *Helops* is usually associated with more friable wood in wetter areas, perhaps supporting *Paxillus panuoides*.

The larvae are long and slender, 25 mm (1 in) or more in length and cylindrical, with a tough skin, typical of this 'click' beetle family. The last segment is equipped with a pair of large, powerful spines curved towards the head, whereas the adjacent segment has two small spines curved towards the rear so that the larva is capable of gripping with these appendages. The life cycle appears to be about 2 years, with a short period of pupation and emergence in May to June. At first the adult beetles are brown but later become deep black with a metallic blue tint. These are particularly handsome beetles 12–25 mm (1/2 in–1 in) long and very active fliers on warm nights.

There are a number of other beetles that are of minor importance. In the family Buprestidae *Buprestis aurulenta* is sometimes found in North American softwoods even 25 years or more after conversion of the wood, apparently because drying has retarded development, as is the case with some Longhorn beetles, as previously described. This handsome beetle from western North America, 15–22 mm (5/8 in–7/8 in) long, is a brilliant metallic green except that the margins of the prothorax and elytra are coppery or red. Other Burprestids sometimes cause damage, particularly in the United States, for example the Turpentine borer, *Buprestis apricans*.

The Dermestidae are another family that should be mentioned as the Hide beetle, *Dermestes maculata*, is sometimes reported as boring into wood in order to pupate; there have been several recent reports in England of its causing damage in hen houses, for this beetle frequently infests hen litter. There are many other Dermestids, such as the Carpet beetles, which are frequently confused with wood-borers, as are so many beetles found in domes-

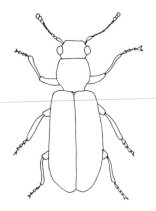

FIGURE B.15 *Korynetes coeruleus*, a predator on Death Watch beetle.

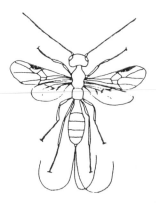

FIGURE B.16 *Eubadizon pallipes*, a predator on Lyctid Powder Post beetles.

tic premises; these are described in more detail in *Remedial Treatments in Buildings* by the present author.

Predators

A more important beetle is *Korynetes coeruleus* (Fig. B.15), a very active metallic blue Clerid 6 mm (1/4 in) long, which is a predator on the Death Watch beetle and whose presence is often an indication of a substantial infestation. The related *Opilio mollis* is associated with the Common Furniture beetle, whilst the smaller *Tarsostenus univitartus* and *Paratillus carus*, blue black in colour except for a white transverse band on the elytra, are often found associated with Lyctid infestations. The larvae, which feed on the wood-borer larvae, are white and have a straight, cylindrical form with a pair of hooks at the posterior end.

Lyctid infestations also attract small Hymenoptera predators, such as the minute, ant-like, wingless Bethylidae, *Sclerodermus domesticus* and *S. macrogaster*, and the minute winged flies of the Braconidae, *Eubadizon pallipes* (Fig. B.16). Other Hymenoptera are associated with the Common Furniture beetle, for example the small wingless, ant-like Chalcydidae, *Theocolax formiciformis* (Fig. B.17) and *Spathius exarata*. All wood-boring beetle infestations also attract the very minute mite *Peymotes ventricosus*;

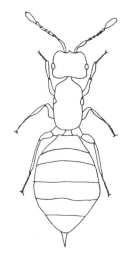

FIGURE B.17 *Theocolax formiciformis*, a predator on Common Furniture beetle.

the female becomes permanently attached to the host larva by its mouth parts, swelling to form a ball about 1 mm (1/24 in) in diameter, whilst the male lives on the body of the female.

Termites

Although various wood-boring beetle infestations occur throughout the world the damage caused by termites is generally far more serious where these insects occur in tropical and subtropical areas. Although termites are sometimes known as white ants they are, in fact, members of the order Isoptera, whereas the true ants are Hymenoptera. However, they have many similarities with ants as they are social insects forming communities which include the male and female reproductive individuals as well as specialized sterile forms or 'castes', the workers and soldiers. Generally the workers are soft-bodied and wingless, confined to the ground or wood, where they devote their energy to feeding, foraging and building. Soldiers serve a defensive role alone and are equipped with large heads and jaws. In the Dry Wood termites of the Kalotermitidae there are no true workers but nymphs instead. A female queen may produce thousands of eggs each day, and the manner of feeding after hatching influences the ultimate differentiation of the forms. Reproductive forms are produced at certain times of the year and disperse to found new colonies.

About 2000 species of termites have been identified, of which more than 150 are known to damage wood in buildings and other structures. Termites are principally tropical but extend into Australia and New Zealand and are common in North America, though rare in Canada (Fig. B.18). Their introduction into certain parts of France and Germany is clearly

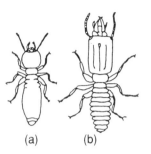

FIGURE B.19 Worker (a) and soldier (b), the most common castes of the termite *Reticulitermes santonensis*.

related to trade; for example the termite of Saintonge, *Reticulitermes santonensis* (Fig. B.19), which was established on the west coast of France between the rivers Garonne and Loire, is now found in Paris around the Austerlitz station, which serves this region. Similarly, *Reticulitermes flavipes* from the United States is concentrated around Hamburg. Neither species can spread widely as they are evidently sensitive to temperature and tend to survive in central heating ducts and other permanently warm areas in buildings.

A common feature of the six families of termites is the lack of a cellulase digestive enzyme, although all these families include wood-

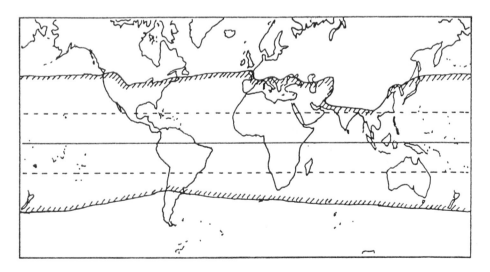

FIGURE B.18 World distribution of termites (after W.V. Harris).

destroyers that are dependent upon intestinal symbionts or prior fungal decay in order to digest cellulose. Identification is difficult in view of the large number of species, yet some classification is necessary in order to decide whether a particular species represents a significant risk justifying some form of preservation. It is also necessary to have a knowledge of the areas in which wood-destroying termites occur, again in order to decide whether preservation treatment is necessary; Table B.1 attempts to summarize this situation.

The damage by the Dry Wood termites of the Kalotermitidae is similar to that caused by the larger wood-boring beetle larvae, such as the House Longhorn beetle, *Hylotrupes bajulus*. In some areas, such as South Africa, both these insects are found infesting wood but the termite damage can be distinguished as the lining of the galleries is smooth, and the galleries are relatively large with distinct, small-diameter connecting tunnels. The faecal pellets are small and cylindrical with rounded or pointed ends but distinct grooves down the sides. In fact, as the Dry Wood termites spread through flying, egg-laying females, as does the House Longhorn

beetle, only the use of naturally durable or adequately preserved wood will avoid damage.

Almost all other wood-destroying termites are subterranean, forming nest cavities in soil or very rotten wood, or mound-building, most constructing covered walkways and all controllable by poisoning the soil of the foundations of a building and a surrounding zone. Termite shields have been widely used in supported buildings to isolate the wooden parts of the structure from the soil but it is difficult to construct shields that are completely reliable and most termites are able to find ways round by constructing mounds or covered walkways; perhaps the main value of the shields is to make the tubular walkways clearly apparent so that they can de destroyed during regular inspections.

European termites

There are only three species of termites which can be described as typically and exclusively European. *Kalotermes flavicollis* occurs throughout the Mediterranean and Black Sea areas, usually in old stumps and dying trees. It has also been reported in vine stock but it appears

TABLE B.1 Wood-destroying termites

Family and genus (number of known wood-destroying species)		Distribution	Notes
Mastotermitidae			Primitive termites, only one known species. Very destructive to wood but in sparsely populated area.
Mastotermes	(1)	North Australia	
Termopsidae			Damp-wood termites.
Zootermopsis	(2)	West United States	
Stolotermes	(1)	New Zealand	
Hodotermitidae			Harvester or forager ants, subterranean, compete with cattle for grass.
Hodotermes	(1)	South Africa	Damage paper, cotton and thatch but not wood; however tunnels and nests damage walls and foundations.
Microhodotermes	(1)	South Africa	
Anacanthotermes	(3)	North Africa, Middle East, India	Can damage softwoods.

Family and genus (number of known wood-destroying species)		Distribution	Notes
Termitidae			Include both subterranean and mound species.
Microcerotermes	(8)	Tropical Africa, Middle East, India, Australia, Central America	
Amitermes	(9)	Tropical Africa, Middle East, India, Australia, Central America	Termites in this group have no intestinal symbiont and can attack only rotted wood as they rely upon fungus to convert cellulose to a digestible form. Only Brown rot fungi are generally involved and lignin is unaffected by either fungus or insect so that it is excreted and used for the construction of honeycomb nests and covered walkways.
Nasutitermes	(13)	Central and South America, Australia, southeast Asia, Ceylon, Mauritius	
Subulitermes	(1)	Central America	
Eremotermes	(1)	Arabia	
Globitermes	(1)	Southeast Asia	
Angulitermes	(1)	Pakistan	
Macrotermes	(4)	Tropical Africa, southeast Asia	Termites in this group are not restricted to rotten wood but instead convert cellulose to a digestible form in fungal gardens within their nests. Fresh wood is gnawed into fragments, chewed and excreted, then infected by White rot fungi, which attacks both cellulose and lignin.
Odontotermes	(15)	Tropical Africa, southeast Asia, India	
Microtermes	(5)	Tropical Africa, Pakistan	
Allodontermes	(2)	South and west Africa	
Ancistrotermes	(1)	West Africa	
Kalotermitidae			Dry-wood termites.
Cryptotermes	(11)	Most tropical countries	The most important wood-destroyers in the tropics able to attack wood at low moisture contents and often known as the Powder Post termites. Intestinal symbionts enable cellulose to be digested but unaffected lignin is excreted. Nests and tunnels are concealed within sound wood and attack is spread by winged, egg-laying females. It is essential to use preserved or naturally durable wood in order to avoid damage, which is similar to that made by House Longhorn beetles.
Neotermes	(4)	South America, Caribbean, India, Seychelles	
Bifiditermes	(1)	Uganda	
Incisitermes	(7)	North, Central and South America, Fiji	
Marginitermes	(1)	North America	
Postelectrotermes	(1)	Madeira	
Kalotermes	(3)	Southern Europe, Middle East, New Zealand	
Rhinotermitidae			Moist-wood termites (cf. Termopsidae).
Coptotermes	(22)	Tropical countries, South America	
Reticulitermes	(10)	North and Central America, Europe, China, Japan	These termites possess protozoan intestinal symbionts but prefer moist wood already infected by fungi or bacteria in order to achieve more effective digestion. They built subterranean nests in soil adjacent to buildings, fences, tree stumps, etc, which they attack through covered walkways if the wood possesses an adequate, constant moisture content.
Heterotermes	(9)	Northeast Africa, Middle East, India, Madagascar, North and Central America, St Helena	
Prorhinotermes	(3)	South America, Caribbean, Guam, Philippines, Solomon Islands	
Psammotermes	(3)	North and South Africa, Middle East, Madagascar	
Schedorhinotermes	(6)	Tropical Africa, southeast Asia, Australia, South Pacific	

that it is actually occupying galleries bored by the Longhorn beetle *Chlorophorus varius*; termites are seldom a primary cause of damage in living plants. This species does not appear to be a true Dry Wood termite in the sense that it is not usually found in dry wood but is apparently dependent upon at least incipient decay.

Reticulitermes lucifugus is found in the same area, approximately south of the Gironde river in France, but there appear to be several races or strains of this; the Iberian, French, Sicilian and Balkan insects are all distinctly different in behaviour. However, they are all subterranean termites which cause extensive damage to interior and exterior woodwork, and they are perhaps the most serious wood-destroying pests in southern Europe.

Reticulitermes santonensis was originally classified as a variety of *R. lucifugus* but it is now considered to be a true species, distinctly more active and more resistant to adverse dry and cold conditions. It is found attacking wood in the open and in buildings in the western coastal area of France between the rivers Garonne and Loire, and has spread along the connecting railway to Paris where infestations are now notifiable. In 1956 it was confined to three areas, with only 55 buildings being affected, but by 1965 86 buildings were known to be infested. Two years later an additional area was discovered and the infested buildings increased to 120, and by 1972 there were a total of 384 known infested buildings and this species was clearly becoming firmly established as a serious wood-destroying pest. This species is also found in Yugoslavia.

Reticulitermes flavipes has been identified around Hamburg and in Vienna but the infestations are very confined in extent and apparently introduced from the eastern United States, where this species, *R. virginicus* and *R. hageni* are the three major subterranean termites. In fact, it is generally true to say that the subterranean termites represent the major problem in the warmer temperate areas, while Dry Wood

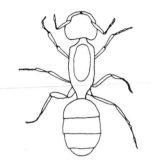

FIGURE B.20 Worker of Carpenter ant, *Camponotus herculeanus*; reproductive castes are winged.

termites, particularly *Cryptotermes* spp, form the major risk in most tropical areas, and the *Kalotermes* spp, perhaps not true Dry Wood termites, are prominent in sub-tropical areas.

Carpenter ants

In many respects the insects of the order Hymenoptera are becoming of increasing significance as wood-borers. The Carpenter ants, *Camponotus herculeanus* (Fig. B.20) and *C. ligniperda*, have been causing increasing damage in buildings in Scandinavia in recent years. In nature these insects tunnel into old trees affected by interior decay in order to establish nests. Modern forestry, however, leaves only very few suitable trees and stumps, and the search for suitable nesting sites probably explains the increasing incidence of infestations in buildings. Summer homes are much more frequently attacked than permanent homes, probably because they are often situated within or close to the forest.

It has been reported that Carpenter ants will attack only wood which has already decayed but this certainly appears to have been discounted in recent years. It has also been suggested that they do not swallow or digest the wood in which they are boring and that they are therefore able to damage wood treated with preservatives containing stomach insecticides, such as copper–chromium–arsenic formulations. The probable explanation for these conflicting reports is the habit of describing both species collectively whereas they are, in fact, distinctly different in behaviour; it certainly appears that *C. ligniperda* is able to attack and utilize dry wood. Damage is typically internal, being an irregular cavity in soft, decayed wood, but it is laminar and follows the growth rings in sound wood.

Camponotus vagus is occasionally found in southern Europe and *C. pennsylvanicus* in North America. Other American species are *Lasius brunneus*, which causes similar damage to *Camponotus*, and *L. fuliginosus*, which attacks wood to obtain material for constructing its nest. There is an increasing danger that these insects will be introduced to other countries in wood shipments and there are already reports of isolated infestations in the British Isles.

Carpenter bees

The order Hymenoptera also includes the family Xylocopidae, the Carpenter bees. These are the largest known bees, black with dark and often iridescent wings, the fine hairs over the body frequently being yellow, white or brown. They generally resemble the Bumble bees, *Bombus* spp, but they are more flattened and less hairy. They are widely distributed in tropical and subtropical areas, but four species have also been reported in France, where the adults have been found boring in beams, rafters and other structural timbers; these bees penetrate very deeply and divide their burrows into a series of cells using fragments of wood. A single egg is deposited in each cell, together with a supply of pollen, and the purpose of the boring appears to be to provide a completely safe egg-laying site. *Hesperophanes cinereus* and *Xylocopa violacea* are frequently reported as causing limited structural damage, principally in central France and along the valley of the Loire but also as far north as Paris; however, the damage is of very limited economic importance.

Saw files

Of similar limited importance is the family Cephidae, the Saw flies, slender-bodied, flying insects which frequently case damage to standing crops. However, one species, *Ametastegia glabrata*, is occasionally found boring in wood and has recently been reported to have caused damage in England to creosote and copper–chromium–arsenic salt-treated posts in motorway fencing. The attack takes the form of a circular entry hole about 3 mm (1/8 in) in diameter leading to unbranched blind tunnels which can be 30 mm ($1^1/_4$ in) in length.

Wood wasps

The most important Hymenoptera are the Siricidae, the Wood wasps. The females use their ovipositers to bore into the bark, laying their eggs below this in the phloem. The latter then becomes infected with a fungus, such as *Stereum sanguinolentuns*, on which the larvae feed. The larvae of *Sirex noctilio*, which are often found in Scots pine, eventually reach a length of 25 mm (1 in) or more before pupating. In the meantime they will have formed an extensive series of galleries which tend to loosen the bark – an important stage in the destruction of sickly and dead trees in the forest. Adults may emerge from wood in service; however, their galleries can be readily distinguished from those of the Longhorn beetles as the flight holes are circular.

Urocerus (Sirex) gigas, the Giant Wood wasp, is found in Scots pine, larch, spruce and fir, producing tunnels up to 6–9 mm (1/4 in–3/8 in) in diameter tightly packed with bore dust. The eggs may be laid as much as 25 mm (1 in) below the surface and the galleries may penetrate into the heartwood, but only sickly trees or felled logs are attacked. The adults, up to 50 mm (2 in) long, are striped yellow and black and often confused with hornets. Trees killed by forest fires can attract great numbers of these insects; they have also been reported as boring in rafters, but it seems most likely that they are introduced in the wood in the forest. Similar damage is caused by other species, for example the blue *Sirex cyaneus*, which infests larch, and also *S. juvencus* and *Xeris spectrum*.

None of these Wood wasps are considered to be of economic significance in Europe as they attack only sickly or dead trees in the forest. However, it appears that they have been introduced into New Zealand and Tasmania, where they are considered to cause damage to healthy trees. As a result the Australian authorities introduced quarantine regulations, these originally being confined to European softwood, which was required to be treated to ensure that it did not introduce further infestations. The regulations have now been extended to all wood imported to Australia, so that all wood shipments must be inspected and all wooden components in packaging and containers must be treated by approved methods, as it has been appreciated that there are many other far more serious wood-destroying pests that could be introduced.

The only other Hymenoptera that are of significance in wood preservation are the predators on Coleoptera, the beetles; these have already been described as their significance lies in the fact that their presence indicates the existence of a substantial infestation by a wood-boring beetle.

Wood-boring moths

The only other insect group of significance is the order Lepidoptera, the butterflies and moths, as two families have wood-boring caterpillars. The Sesiidae, the Clear wings, are not readily identifiable as types of moths as they more closely resemble wasps or bees in that they have rather slender bodies; however, they can be identified by the narrow band of scales around the edges of the wings. The Cossidae are the Goat moths; *Cossus cossus (ligniperda)* can be found boring into the base of oak, elm, willow and poplar trees. The larvae bore large galleries, frequently causing serious damage to standing trees, principally because many of the species are very large, some Australian adults achieving a wing span of 180 mm (7 in). In Europe and North America the most important species

is the Wood Leopard moth, *Zeuzera pyrina (coesculi)*, which bores in upper branches, usually in fruit trees. This species has a wing span of about 45 mm (1³/₄ in) and is generally considered to be an occasional serious pest of fruit trees, although it has sometimes been suggested that it may be the cause of damage in structural wood.

The family Pyralidae should also be mentioned for, although these social moths are not wood-borers, their webs and pupal chambers are often found in deep, open joints in old roof timbers, where they are sometimes confused with fungal growth by inexperienced surveyors. The two most important species are the Bee moth, *Aphomia sociella*, and the Honeycomb moth, *Galleria mellonella*.

Marine borers – Gribble

The most important marine borers are Crustacea of the sub-order Isoposa. They are all superficially similar to the common wood louse, which is also included in this sub-order. *Limnoria* spp, the gribbles, are 3–5 mm (1/8 in–1/5 in) long and strong swimmers; however water currents generally have a greater influence over their distribution. They settle on wood, forming superficial burrows less than 12 mm (1/2 in) deep with small entrance holes of less than 2.5 mm (1/10 in) diameter. The galleries follow the early wood, giving it a laminar appearance, until the weakened zone breaks away, exposing a fresh surface to attack. Gribble actually attacks wood at all depths below the mid-tide level but damage is most apparent in the tidal zone where the wave action steadily removes damaged wood, permitting progressive erosion to occur. Eggs are held by the female in a brood pouch beneath the body until they eventually hatch and the young gribble are released into the parent's burrow. These eventually bore on their own and also swim freely in warm weather in search of suitable settlement sites.

Various species of *Limnoria* occur throughout the world, and in tropical areas *Sphaeroma* spp, which are similar in habits but about 15 mm (3/5 in) long, are also found, as well as the related *Exosphaeroma* spp. *Chelura* spp, members of the sub-order Amphipoda, may also be found infesting wood; however, they do not appear to cause damage, but simply occupy vacant gribble tunnels.

Shipworm

The molluscan borers are all included within the order Eulamelli branchiata, but are in two families. The Teredinidae include pile worms, shipworms and cobra. The best known genus is *Teredo*, the shipworms. The minute larvae swim freely and eventually settle on wood where they bore holes about 0.5 mm (1/50 in) in diameter. Throughout the rest of its life the *Teredo* remains within the wood, the only visible evidence of its presence being two syphons projecting from the burrow, which enable water to be drawn in and discharged. If the burrow becomes exposed to the air, perhaps during low tide, the syphons are withdrawn and the hole sealed with the pallet. The *Teredo* grows, and extends its tunnel in order to accommodate its enlarging body by using its shells as cutters, boring principally along the grain. The burrows have a characteristic calcareous lining but the extensive damage that may be caused in suitable conditions is almost entirely concealed, unless it is exposed by gribble damage causing the surface to break away.

Various species of *Teredo* are widely distributed throughout the world, although confined to saline waters which are reasonably warm; in Europe infestations are generally confined to southern and western coasts exposed to the Gulf stream. *Bankia* spp are generally far larger than *Teredo* and are confined to the tropics.

The family Pholadidae, the boring mussels, is of minor importance as the damage caused is relatively superficial. *Martesia* is sometimes found boring in wood, usually after this has been softened by decay, but other species are usually found boring in stone or soft mud.

The purpose of this appendix has been to illustrate the wide variety of borers that exist so that the reader can be aware of the need for preservation. The identification of adult insects can be attempted by reference to the descriptions given in the text and the figures, but the identification of borer damage is rather more complex and outside the scope of this book, which is concerned simply with the preservation of wood in order to prevent attack; the identification of damage is considered in more detail in the book *Remedial Treatments in Buildings* by the present author.

Appendix C

Wood-destroying fungi

Sapstain

Damp wood is able to support a wide variety of fungal infections. In fresh commercial wood sapstain is the first defect that may occur, Ascomycetes and Fungi Imperfecti developing in the residual moisture of the tree, where both a high moisture content and the presence of sugars are ensured. Sapstain, which, as its name implies, is generally confined to sapwood, results from deeply penetrating fungi which cause discolouration through their dark-coloured hyphae or occasionally through staining of the cell walls. The discolouration may be black, green, purple, pink or, very occasionally, brown, but it is most commonly greyish-blue, particularly on softwoods, and is frequently described as 'bluestain'. The hyphae that cause this bluestain can, in fact, be seen to be dark brown when examined microscopically and the blue colouration results from refraction of incident light by these hyphae. The stained areas tend to follow porous routes, spreading along the grain and radially to form patches which are wedge-shaped in cross-section.

Most bluestaining appears to be caused by Ascomycetes of the genus *Ceratocystis (Ceratostomella)* but a number of Fungi Imperfecti also cause staining in coniferous wood, the most important being *Aureobasidium (Pullularia) pullulans, Hormiscium gelatinosum, Cladosporium herbarum, Cadophora fastigiata, Diplodia* spp and *Graphium* spp. Sapstain in Scots pine or European redwood appears to be

due, in decreasing order of importance, to *Ceratocystis pilifera, C. coerulescens, C. piceae, Aureobasidium pullulans* and *C. minor*. These fungi accounted for 90% of bluestain hyphae isolated from pine in investigations by Professor Henningsson is Sweden. In spruce the order was slightly different, but the same principal fungi were involved: *Ceratocystis piceae, C. coerulescens, C. pilifera* and *Aureobasidium pullulans*.

Mould

Sapstain is almost invariably associated with superficial discolouration caused by moulds forming greenish or black, occasionally yellow, powdery growths, which are easily brushed or planed away. A very wide range of species is able to develop in this way on damp surfaces of wood, these species including common genera such as *Penicillium, Aspergillus* and *Trichoderma*, none of which causes significant deterioration or deeply penetrating stain. These mould growths occur only when the wood is freshly felled as dampness and sugars are then both available.

The superficial darkening of weathered wood is usually caused by *Aureobasidium (Pullularia) pullulans, Cladosporium herbarum, Alternaria* spp or *Stemphylium* spp, all of which develop minute dark pustules. *Aureobasidium pullulans* does not confine its activities to exposed wooden surfaces but is also commonly isolated from painted or varnished surfaces, together with

Phoma spp. It is often said that this growth results from applying paints or varnishes on top of sapstained wood but this cannot explain how these fungi also occur when finishes are applied to metal surfaces. In fact, it appears that these fungi are able to attack the coating, perhaps forming bore holes and spreading into the wood beneath, where they can then cause 'stain in service' in contrast with the better known 'sapstain on freshly felled wood'.

Bacteria

Generally, the moulds and staining fungi have no significant effect on wood, except that the staining fungi, in common with some bacteria, are able to utilize cell contents, particularly materials tending to block the pits, so that their activities result in a distinct increase in porosity, particularly in the sapwood of impermeable species such as spruce. Some of the Ascomycetes and Fungi Imperfecti can, however, cause damage to the cell walls, resulting in a form of deterioration commonly known as Soft rot.

Soft rot

The relationship between this damage and fungi was established only comparatively recently by Mr Savory in England and new species of Soft rotting fungi are being progressively identified. At the present time Soft rot is considered to be associated with wood at very high moisture contents, perhaps saturated but in aerobic conditions, such as in water-cooling towers, where damage of this type was first identified. The deterioration takes the form of surface softening which becomes progressively deeper, small cuboid cracks developing if the affected wood is dried. If decay becomes very deep the wood breaks with a distinct brash or cross-grain fracture. It seems likely that Soft rots may be responsible for unexplained brashness in wood which is apparently unaffected by fungal attacks.

Soft rotting fungi are often termed microfungi as they are able to progress through wood within the cell walls, where they are not readily identified, as compared with the Basidiomycetes, which normally progress through the cell lumen and pits. As some preservative toxicants are deposited on the lumen surfaces of the cell walls it is hardly surprising that many Soft rots are very resistant to such preservative systems and are readily controlled only by toxicants that penetrate into the cell-wall structure. Although this preservative tolerance is insignificant in softwoods it is a matter of great concern in hardwoods, particularly tropical species, in which Soft rot may progress in ground contact conditions, despite very high retentions of otherwise effective preservatives such as creosote and copper–chromium–arsenic salts.

Basidiomycete classification

The principal wood-destroying fungi are Basidiomycetes, the spores of which are borne on small, club-shaped structures known as basidia, which are normally formed in a compact layer called the hymenium. The wood-destroying fungi, Hymenomycetes, consist of four families, which differ in the form of the hymenium. The Thelephoraceae, a family that includes *Coniophora*, have the hymenium freely exposed on a flat, skin-like surface. In the Hydnaceae family the hymenium is on a surface of spine-like outgrowths, whilst that of the Polyporaceae family, which includes *Fomes*, *Lenzites*, *Serpula*, *Polyporus*, *Coriclis*, *Poria* and *Trametes*, lines the inside of pores or tubes. The Agaricaceae, which include *Lentinus* and *Paxillus*, are quite distinctive as the hymenium is on plate-like gills underneath a cap-shaped pileus (or mushroom).

Brown and White rots

If a wood-destroying fungus is producing spores it is possible to identify its family, but in the wood-preservation industry it is usually necessary to rely upon the nature of the decay and the superficial appearance of the hyphae, as sporelation is comparatively rare in most important species. One of the most useful identi-

FIGURE C.1 Advanced decay by Dry rot, *Serpula lacrymans*, showing typical cuboidal cracking and fungal growth. (Cementone Beaver Ltd)

fication features is the division of these Basidiomycetes into Brown rots and White rots, depending upon the manner in which they destroy wood.

In a Brown rot, such as *Coniophora puteana*, the fungal enzymes destroy the cellulose but leave the lignin largely unaltered so that the wood acquires a distinct brown colour and the structural strength is almost entirely lost. As decay progresses the wood becomes very dry and shrinkage cracks appear both across and along the grain, the size and shape of the resulting rectangles often being a useful feature in identification. In contrast, the White rots, such as *Coriolus (Polystictus) versicolor*, destroy both cellulose and lignin, leaving the colour of the wood largely unaltered but giving a soft felty or stringy texture.

Most of the wood-destroying fungi are confined to forest situations but those which are described below are often found in structural wood, particularly in buildings, and occur widely throughout the world, although certain species tend to dominate in particular areas.

Dry rot fungus, *Serpula lacrymans*

The best known wood-destroying fungus is certainly *Serpula (Merulius) lacrymans*, the Dry rot fungus (Fig. C.1). This species is usually found in buildings, and sometimes in mines, in places where ventilation is restricted and it thus tends to develop in completely concealed areas. It appears to have originated as a north European species but it is now found in other parts of the world with similar climatic conditions, such as North America and parts of South Africa, Australia and New Zealand. It is comparatively rare in warmer climates, except where it is associated in structures with condensation caused by the air conditioning system.

The conditions for germination and growth

are extremely critical, requiring a narrow range of atmospheric relative humidity and wood moisture content. Spore germination appears to occur most readily in acid conditions and it is these features that perhaps account for the fact that *Serpula lacrymans* is frequently associated with other fungi able to germinate and develop in wetter conditions. If wood becomes accidentally wetted one of the Wet rot fungi, such as *Poria vaporaria*, may develop but as drying progresses conditions may arise in which *Serpula lacrymans* spores may germinate, encouraged by the acid conditions caused by the previous Wet rot. Similarly a source of continuing dampness may support a Wet rot such as *Coniophora puteana* but at a further distance from the source the moisture content may be optimum for *Serpula lacrymans* spore germination, again encouraged by the acidity generated at the fringe of the *Coniophora puteana* attack.

Serpula lacrymans produces white hyphae which are, in fact, very fine tubes or hollow threads, progressively branching and increasing in length so that they spread in all directions from the initial point of germination, provided that a food source is available. As food is exhausted some hyphae are absorbed whilst others are developed into much larger rhizomorphs or conducting strands, which are able to transport food and water. Thus exploring hyphae finding no nourishment are absorbed to form food for growth in more promising directions, giving the fungus the appearance of sensing the direction in which to spread towards a food source. Seasonal changes sometimes inhibit growth, which then resumes when suitable conditions return. In this way hyphae contract on drying to form layers or mycelium, each successive layer indicating a season of growth.

Active growth is indicated by hyphae like cotton wood, perhaps covered with 'tears' or water drops in unventilated conditions, this being the way the fungus regulates the atmospheric relative humidity and accounting for the name of 'lacrymans'. Rhizomorphs may be up to 6 mm (1/4 in) in diameter, they are relatively brittle when dry and extend for considerable distances over and through brickwork, masonry and behind plaster, spreading through walls between adjacent buildings and ensuring a residue of infection, even if all decayed wood is removed; treatment of adjacent sound wood and replacement of decayed wood with preserved wood should always be accompanied by sterilization treatment of infected brickwork and masonry.

Mycelium is greyish and later yellowish with lilac tinges when exposed to light, and often subsequently green in colour through the development of mould growth. Sporophores generally develop when the fungus is under stress through the food supply being exhausted, the temperature increasing or the moisture content decreasing. Sporophores are shaped like flat plates or brackets and vary from a few centimetres to a metre or more across, being grey at first with a surrounding white margin but then the slightly corrugated hymenium or spore-bearing surface develops to become covered in millions of rust-red spores which are eventually liberated and cover the surroundings with red dust. As fungal growth in buildings is generally concealed the sporophore may be the first sign of damage, though a characteristic mushroom smell may be noticed if an infected building is closed for several days.

Serpula lacrymans can cause severe Brown rot with pronounced cuboidal cracking, the cubes being up to 50 mm (2 in) along and across the grain. Decayed wood crumbles easily between the fingers to a soft powder. Two important features of the decay are the fact that it can be entirely internal and concealed in beams, and that it can spread to dry wood in unventilated conditions, as the fungus is able to transport adequate water for decay through the rhizomorphs. A related species, *Merulius himantioides*, is sometimes found causing similar decay in Scotland, Denmark and southern Sweden.

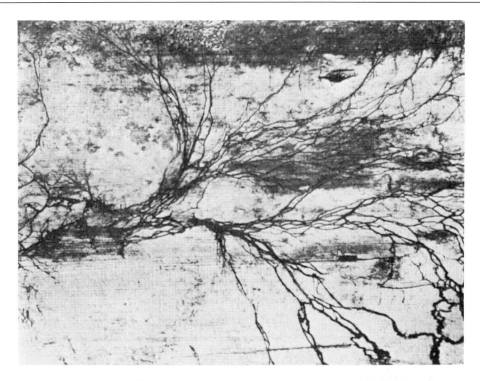

FIGURE C.2 Advanced decay by the Wet rot, *Coniophora puteana*, showing dark fungal strands. (Cementone Beaver Ltd)

Cellar fungus, *Coniophora puteana*

The Cellar rot fungus, *Coniophora puteana* (*cerebella*) (Fig. C.2), is the most common cause of Wet rot in buildings and elsewhere where persistently damp conditions arise through, for example, soil moisture or plumbing leaks. Spores germinate readily and this fungus is likely to occur whenever suitable conditions arise. The hyphae are initially white, but growth is not as generous as for *Serpula lacrymans*. In addition there is little development of mycelium and only thin rhizomorphs are formed. These rhizomorphs, which become brown and eventually black, are not so extensive as those of *Serpula lacrymans* and never extend far from the wood. The sporophore occurs only rarely in buildings and consists of a thin skin covered with small irregular lumps. The hymenium is initially yellow but darkens to olive and then brown as the spores mature. Wood in contact with a source of moisture such as brickwork often consists of a thin surface film concealing extensive internal decay. The rotted wood is dark brown with dominant longitudinal cracks and infrequent cross-grain cracks. The easiest method for controlling *Coniophora cerebella* in buildings is to isolate wood from the source of dampness.

White Pore fungus, *Poria placenta*

The White Pore fungus, *Poria placenta* (Fig. C.3) is common in mines and occasionally occurs in buildings. It requires a higher moisture content than *Serpula lacrymans* but, in contrast to *Coniophora puteana*, it is tolerant to occasional drying and is therefore the normal fungus associated with roof leaks. Growth is generally similar to *Serpula lacrymans* but strands remain white, compared with yellow

215

FIGURE C.3 Strands of a wet rot, a *Poria* species. (Cementore Beaver Ltd)

or lilac for *Serpula* and brown or black for *Coniophora*. The rhizomorphs may be up to 3 mm (1/8 in) in diameter but are not so well developed as those of *Serpula* and are flexible when dry. When examined on the surface of a piece of wood or adjacent masonry they appear to be distinctly flattened or to have a flat margin on either side of the strand, and they do not extend far from their source of wood. The sporophore is rare in buildings but it is sometimes observed in greenhouses when severe decay occurs. It is a white irregular plate, 1.5–12 mm (1/16 in–1/2 in) thick and covered with distinct pores, sometimes with strands emerging from its margins. The decay damage to wood is similar to that caused by *Serpula* but the cubing is somewhat smaller and less deep. When decayed wood is crumbled between the fingers it is not so powdery as that attacked by *Serpula* but slightly more fibrous or gritty.

Other *Poria* species – dote

Poria xantha is a similar Brown rot, being frequently found in greenhouses but usually with no visible surface growth, although mycelium may be found in cracks, even in the cubing cracks on decayed wood. A thin skin of yellow- ish white mycelium occasionally occurs. The sporophore is a thin, yellowish layer of pores, distinctly lumpy when situated on a vertical surface. *Poria monticola* is sometimes found both in buildings and as dote on softwood imported from North America, where it is one of the most important wood-destroying fungi. Dote is a form of pocket rot which can develop when unseasoned wood is close stacked and it is not easy to detect. It may, however, be visible as faint streaks or elongated patches of yellowish or pinkish-brown on most softwoods, or of a purplish colour on Douglas fir. When tested with the point of a knife the wood is found to be brash within these patches. As decay progresses the wood acquires a typical brown colour with cracks along and across the grain, as when attacked by the other *Poria* spp.

Stringy Oak rot, *Phellinus megaloporus*

The Stringy Oak rot, *Phellinus megaloporus (cryptarum)*, occurs in Europe on oak in conditions in which *Coniophora puteana* or *Poria placenta* is normally found on softwoods, such as in association with roof leaks or masonry affected by soil moisture and it is able to resist the relatively high temperatures that frequently occur in roof spaces. As this fungus prefers oak it is largely associated with older buildings constructed with this wood. It is a white rot causing no distinct colour change in the decayed wood, but this becomes much softer, with a longitudinal fibrous texture, and does not powder in the same way as wood decayed by *Serpula* or other Brown rots. Yellow or brown mycelium is sometimes formed on the surface of wood. The sporophore is a thick, tough plate or bracket, fawn coloured but darkening as the spores develop. *Poria medulla-panis* causes very similar decay in oak, particularly in timber-framed buildings where oak is exposed to the weather, but there is no practical reason in terms of preservation or treatment why these two species should be differentiated.

Coriolus versicolor

Coriolus (Polystictus) versicolor is the commonest cause of White rot in hardwoods, especially in ground contact, but is usually confined to the sapwood in durable species. It causes decay on hardwood props in mines and has been occasionally reported as causing decay in sapwood of softwoods, particularly in external, painted joinery (millwork) such as window and door frames. The sporophore is rarely seen but consists of a thin bracket up to 75 mm (3 in) across, grey and brown on top with concentric, hairy zones and a cream pore surface underneath from which the spores are released. Infected wood initially suffers white flecking and eventually bleaches in colour. Shrinkage is rare and the decayed wood simply appears to be lighter and much weaker than sound wood.

Lentinus lepideus

Lentinus lepideus (squamosus) is the principal decay found in railway sleepers (ties) and transmission poles, perhaps because it is comparatively tolerant to creosote and can therefore develop when treatment with this preservative is inadequate. A brown cuboidal rot is caused

and the white mycelium, perhaps with brown or purple tinges, can often be observed within the shrinkage cracks. The sporophore is on a stem, woody and brownish with a gill extending down the stem when it develops in normal light. In limited light the form is abnormal, perhaps lacking the cap or having branched, cylindrical, white or purplish-brown outgrowths.

Lenzites sepiaria

Lenzites sepiaria is comparatively rare in Europe but occurs on imported wood which has been infected in the forest and continues to develop if the wood is used for fencing, bridging, poles or situations in buildings where there is an adequate moisture content. The first evidence of infection is a pale yellow zone accompanied by softening and brashness of the fibres. The wood becomes progressively darker brown and slight cuboidal cracking occurs. Superficial hyphae are rarely observed but orange-yellow mycelium may occur on decayed wood in concealed situations. The sporophore is a bracket, up to 50×100 mm (2 in $\times$ 4 in) in size, tough and with a hairy upper surface and distinct gills underneath. It is tawny yellow at first but later dark brown with a yellow margin and brown gills.

Lenzites trabea – dote

Lenzites trabea appears to have originated in North America, but it is now well established in central and southern Europe. It is sometimes found in the British Isles and northern Europe on imported wood, frequently as a dote or pocket rot originating as a decay of softwood in the forest. The rot can develop in suitable conditions, causing a brown cuboidal decay. The sporophore is a thin bracket, tough and with gills on its underside, yellow-brown at first but becoming darker and then bleaching on the upper surface. This species appears to be particularly attractive to some termites, such as *Reticulitermes flavipes*.

Paxillus panuoides

Paxillus panuoides causes decay similar to that caused by *Coniophora cerebella* but tends to occur in much wetter conditions. The hyphae develop into fine, branching strands which are yellowish, never becoming darker, and a rather fibrous, yellowish mycelium, perhaps with lilac tints, may occur. The wood is stained bright yellow in the early stages, but darkens to a deep reddish-brown and shallow cracking occurs. The sporophore has no distinct stalk but is attached at a particular point, tending to curl around the edges and eventually becoming rather irregular in shape. The branching gills on the upper surface radiate from the point of attachment. The colour is dingy yellow but darkens as the spores develop; the texture is soft and fleshy.

Dote or pocket rot

Dote consists of narrow pockets of incipient decay and is sometimes described as pipe or pocket rot. The rot in the pocket is brown and cuboidal and hyphae like cotton wool may be present. If a pocket is absent, except for the appearance of a brown stain, dote may be confirmed by brashness when the stained area is probed with the point of a knife. Dote is observed usually in softwood imported from North America; *Lenzites* and *Poria* spp have already been described as two of the possible causes. Another is *Trametes serialis*, which is usually a cause of heart rot in standing trees or closely stacked, unseasoned boards. This fungus will continue to develop and the decay will spread if the wood remains damp; in South Africa *Trametes serialis* on imported wood has been able to develop locally and is now as serious as *Coniophora puteana* or *Poria placenta*.

Trametes serialis was at one time confused with *Poria monticola*, which can cause similar dote; this is the way in which infections of this fungus are generally introduced into Europe. *Fomes annosus*, a parasite of sickly trees in which it can cause heart rot, turning the wood into long, brown, fibrous strands, can continue to develop and slowly decay wood after this has been felled if the moisture content is maintained, as in pit props. This species is another form of dote, the decay pocket originating as a black spot and then enlarging progressively and becoming filled with hyphae which have the appearance of white lint.

North American species

The distribution of individual fungi is more extensive than that of wood-borers, although in many cases the fungi evidently originated in one particular country and spread to other areas with wood exports. Certainly the principal wood-destroying fungi in buildings in North America have a distinctly different balance to those found in Europe, but it is clear that trade in both directions is progressively exchanging species. At the present time *Coniophora puteana* is as important in North America as in Europe. *Serpula lacrymans* is confined to the northern United States and Canada, but *Poria incrassata*, which is similar in many ways and which causes similar damage, is found only in the southern United States. *Coniophora arida* is sometimes found causing decay on preserved pine and *Lentinus lepideus* is the major decay fungus on wood treated with creosote, as in Europe. The *Lenzites* spp are perhaps most prevalent of all, with *L. trabea* commonly occurring on both softwood and hardwood in service. *Poria monticola* is also a serious cause of decay of wood in service, but it appears that infection is probably always introduced in standing trees or felled logs in the forest.

This appendix has attempted to review the fungi that are the most serious causes of decay in wood in service. However, it will be appreciated that there are many other forest fungi which may occasionally be encountered in buildings, though they are not necessarily associated with wood decay; these latter fungi are described in more detail in *Remedial Treatments in Buildings* by the present author.

Index

Index